The Media Creates Us in Its Image

and Other Essays on Technology and Culture

Richard Stivers

CASCADE *Books* · Eugene, Oregon

THE MEDIA CREATES US IN ITS IMAGE AND OTHER ESSAYS
ON TECHNOLOGY AND CULTURE

Cascade Books
An Imprint of Wipf and Stock Publishers
199 W. 8th Ave., Suite 3
Eugene, OR 97401

www.wipfandstock.com

PAPERBACK ISBN: 978-1-5326-9725-8
HARDCOVER ISBN: 978-1-5326-9726-5
EBOOK ISBN: 978-1-5326-9727-2

Cataloguing-in-Publication data:

Names: Stivers, Richard, author.

Title: The media creates us in its image and other essays on technology and
culture / by Richard Stivers.

Description: Eugene, OR: Cascade Books, 2020 | Includes bibliographical refer-
ences.

Identifiers: ISBN 978-1-5326-9725-8 (paperback) | ISBN 978-1-5326-9726-5
(hardcover) | ISBN 978-1-5326-9727-2 (ebook)

Subjects: LCSH: Technology—Social aspects. | Mass media—United States. |
Jacques Ellul.

Classification: HM846 .S70 2020 (print) | HM846 .S70 (ebook)

Manufactured in the U.S.A.   03/27/20

To my wife Janet and all my family and friends.

Contents

Acknowledgments

The life and work of Jacques Ellul have inspired my own work. Jim van der Laan, Kim Goudreau, and Nick Maroules have kept me on the right track. Once again, Sharon Foiles did an excellent job of typing and patiently handling my indecision. A special thanks to Lauren Brown, friend and co-author of "The Legend of 'Nigger' Lake: Place as Scapegoat."

Introduction

The following essays, inspired by the work of Jacques Ellul on technology, were written over the past twenty-five years.[1] In my estimation they are still timely. Modern technology is always changing, yet the goal remains the same—ever greater efficiency. Social media, for instance, are more efficient than the mass media in bringing about conformity, but in both types of media conformity is the paramount social (in contrast to economic) function.

Ellul recognized that technology includes both material and nonmaterial techniques. The latter consist of psychological and organizational techniques, such as advertising and bureaucracy. The exploitation of our physical environment demands the increased control of humans in the interest of adjustment to technological change. Maladjusted humans are simply a problem of inefficiency.

In a technological civilization, Ellul argued, everything becomes an imitation of technology or a compensation for its impact on society. Technology thus supplants culture in becoming the chief determining factor in the organization of modern society. The progress of technology necessitates it becoming a system. Concurrently, however, technology is the primary disorganizing force in society by crowding out culture, making it fragmented and ephemeral. Culture simply becomes compensation and play: Meaning and truth are reduced to a game. The fragmentation of culture is paralleled by the fragmentation of the moral self—we become mere role players with no stable identity.

The following essays cover a broad range of topics from morality, sin, and ritual scapegoating to personality and male immaturity. Each essay highlights technology either as an organizing or a disorganizing factor or both. The wide range of topics reflects the pervasiveness of technology in society.

1. See especially Ellul, *Technological Society*; Ellul, *Technological System*; Ellul, *Technological Bluff*.

In "Technique Against Culture," I examine how technology concomitantly destroys meaning and creates false meaning in its place. For Christians truth and meaning have a specific reference—the life of Jesus Christ. Nevertheless Christians need to work with non-Christians to help bring about a livable society without regarding its cultural values as absolute. It is possible, then, to distinguish between "genuine" and "spurious" cultures.[2] A genuine culture integrates the lived experiences of people so that work, play, and human relationships form a whole giving even suffering and death a purpose. A spurious culture is a culture of compensation that is imposed on people. It permits and even encourages human relationships to be exploitative, while only providing compensation. Our advertising culture, for example, creates a false meaning of consumerism as compensation for the violent nature of our economic, political, and social relationships. It promotes a hopeless live-for-the-moment creed. This culture of false meaning is likewise a fragmented culture of constant experimentation.

"The Media Creates Us in Its Image" argues that the media (mass and social) function to create a human being that is the opposite of the ideal in Western civilization since the Middle Ages. Greek, Roman, and Christian ideals, even though not compatible, were part of a grand synthesis that placed value on the individual, reason, and freedom (with responsibility). In contrast, today we have internalized and normalized technological stimuli so that we demand ever greater speed, noise, and information. Both the form and the content of the media work to create adults with at least mild cases of hyperactivity and attention deficit disorder. The media encourage distraction, shallow relationships, loneliness, need for instant gratification, and an addiction to the peer group. In short, the media give birth to children, who Peter Pan-like never grow up. "We have become the childish machines we created: The media creates us in its image."

A technological society gives rise to a technological personality. I describe this contradictory personality in "The Technological Personality." In the late eighteenth century, science and technology gradually supplanted religion as the basis of truth. Religion and eventually morality began to be regarded as subjective. With the decline of a moral unity in society, human relationships became vague, uncertain, and even dangerous. Loneliness and a fear of others were the result. This left the individual in greater need of the peer group for protection and identity. As David Riesman termed it, the other directed person was anxious about being popular and fitting in.[3] At the same time business was demanding an extroverted employee, who appeared to be

2. Sapir, "Culture, Genuine and Spurious."
3. Riesman, *Lonely Crowd*.

cheerful, friendly, and confident. Such employees were perceived to be more likely to get along with customers and fellow employees. Introverts were under suspicion. The extroverted person was the outer shell of the technological personality, the lonely, anxious individual its inner core.

"No Place for Men" confronts the problem of male immaturity. For over two hundred years the period of emotional and intellectual immaturity for both sexes has lengthened. The main reasons are a lack of moral unity in society and the homogeneity that science and technology produce through quantification. The former means that there is a discontinuity between generations, whereas the latter requires that the future must be based on information and prediction, making the choice of marriage and occupation highly problematic. Prolonged immaturity is the consequence.

Male immaturity is complicated, among other things, by the demand for an equality of power. Left to itself an equality of power as a social goal only leads to competition with no possible resolution. The monopoly and abuse of power by men has led to their castigation. Masculinity itself appears to be equated with abuse of power. If men are bad, some boys may indefinitely postpone adulthood. Boys are escaping into the world of video games and refusing to take on the responsibilities of a tainted manhood.

"The Computer and Education" is a warning to teachers to question any new technology used in the classroom. The computer does not enhance serious education; it only allows students to access information and solutions quicker and with little effort. Its negative functions, however, are serious. The computer reduces words to an abstract and single meaning in removing their context. In objectifying words it turns them into "objects." The computer privileges logical thought over dialectical thought. The latter is essential to the humanities and social sciences.

The computer permits the proliferation of information, all at our disposal. The result is a cynical worldview that information is random and exists to be exploited, not understood and appreciated. Most importantly the computer substitutes quantity of information for moral judgment. The computer thus becomes the teacher and the human its aide.

The modern university serves the technological system and global capitalism. In "The Need for a 'Shadow' University," I describe its current state and an alternative. From the Middle Ages to the present the university has faced numerous challenges to its autonomy: the church, the state, and now the technological system. Every aspect of the modern university—administration, pedagogy, research—is specialized and technical. The curriculum is replete with courses on methodology and statistics. Students are being prepared to serve the technological system. The humanities have no purpose in a society whose sole goal is efficiency. They are at best a distraction.

Serving as a critique of the technical university is an informal "school" in which professors teach (without pay) courses in the humanities and social sciences in particular, courses that both conserve and criticize our cultural heritage. Such courses would raise questions about the meaning of life. No grades, no tests, no degrees would be given. This "shadow" university would be free from the constraints under which the modern university labors.

Since the early nineteenth century modern societies can be characterized in terms of both extreme individualism and extreme collectivism. The individual, cut off from the control and support of extended family and community, had become autonomous, but this individualism was psychologically debilitating; consequently, the individual turned to public opinion and government for guidance.

In "Ethical Individualism and Moral Collectivism in America," I apply this paradox to morality. Another term for ethics is theoretical morality. The latter rarely becomes the lived morality of a society, but often acts as an ideology. Lived morality refers to the effective moral attitudes of a society. Ethical individualism assumes the form of subjective preferences and choice of values: ethical consumerism. Moral collectivism is an ersatz morality consisting of technical and bureaucratic norms, public opinion, and visual images in the media. Moral collectivism permits neither individual freedom nor individual responsibility. Ethical individualism serves as its compensation.

Tradition is primarily oral. The shared symbolic experiences of the past are transmitted from generation to generation. "Visual Morality and Tradition" argues that the visual images of the media, by contrast, are an essential part of an ersatz morality that binds us to material reality. Only language permits us to raise the question of truth and meaning. Visual images are norms of our present reality. What is replaces what ought to be. A technological culture conflates truth and reality, and replaces the tension between what is and what ought to be with that between what is and what is technologically possible.

The Judaic-Christian idea of sin is expressed metaphorically most often as slavery or bondage. In "Sin as Addiction in our 'Brave New World,'" I suggest that sin as addiction is a more relevant metaphor today. Earlier Ellul maintained that the metaphor sin is alienation better fits our condition under industrialized capitalism.[4] To be alienated is to have someone else control one's life.

Sin implies more than immoral conduct; it is a spiritual matter of possession. As Søren Kierkegaard observed, sin is not so much discrete action

---

4. Ellul, *Ethics of Freedom*.

as a condition.[5] Both slavery and alienation then refer to a condition of external possession.

Sin as addiction is an apt metaphor for our life in a technological society. With the help of social media, advertising has found new ways to make us become addicted to the goods, services, and information it offers us. In *Addiction by Design*, Natasha Schüll details how gambling machines are designed to create addiction. The use of behavioral psychology is crucial to this effort. The "fear of missing out" is fundamental to the effort to make us use various social media continuously.

Sin as bondage, alienation, and addiction all point to our spiritual possession by the enemy. All three point to the truth about sin.

"The Festival in Light of the Theory of the Three Milieus: a Critique of Girard's Theory of Ritual Scapegoating" maintains that Rene Girard's theory of ritual scapegoating is not universal: It does not encompass all history. Using Ellul's theory of the three milieus that humans have lived in, I argue that Girard's theory applies principally to the milieu of society. A milieu is our most immediate environment. Each milieu is based on opposite principles: nature, on life and death; society, on good and evil; technology, on efficiency and inefficiency.

Girard's theory claims that scapegoating is the origin of ritual and religion. The scapegoat embodies moral evil. Ritual scapegoating is the "solution" to the violence that mimetic desire engenders. Mimetic desire is desire whose real object is the other person. In desiring to be like the other person, I desire what she desires. This theory, I maintain, only fits the milieu of society.

In the milieu of technology, desire and imitation exist in relation to technological objects and images, including the celebrity (a creation of the media). Moral evil has no place in a technological society; it is reduced to inefficiency.

"The Legend of 'Nigger' Lake: Place as Scapegoat" extends Girard's theory of ritual scapegoating from a person to a place. A wetlands, known locally as "Nigger Lake," was blamed for the flooding problems in a community even though experts told residents at a town meeting that the wetlands had nothing to do with the flooding. The scapegoat is always an outsider, whose removal will restore unity in the community. Some believed that the water in this Illinois wetlands originated in Ohio and even New York. Residents urged the destruction of the wetlands.

Aldous Huxley, to my knowledge, is the only author to have written a work of science fiction, *Brave New World*, and then almost thirty years

5. Kierkegaard, *Sickness Unto Death*.

later a work of nonfiction, *Brave New World Revisited*, that assessed whether his forecasts had come to fruition. They had. In *Brave New World*, Huxley understood that efficiency, order, and conformity were the primary social concerns in a technological society. In the introduction to the 1946 edition of *Brave New World*, he mentioned that "inefficiency is the sin against the Holy Ghost."[6] The novel depicts an array of psychological techniques for the manipulation and control of society. Manipulation was based on the desire for pleasure; by contrast Orwell's *1984* anticipated coercion as the main technique of control. The most important issue Huxley missed was one Ellul discussed in *The Technological Society*: The more rational, logical, and demanding a society becomes, the more the need to rebel spontaneously and irrationally against the system arises. The rebellion, however, is partially absorbed by the system and transformed into techniques of compensation and escape, such as self-help groups.

"Technology and Terrorism in the movie *Brazil*" concludes that terrorism leads to an increase of power in the political state, bureaucracy, and technology. Terrorism would appear to be necessary for the advance of modern organization, so much so that in the movie no one ever sees a real terrorist, only those who violate bureaucratic rules. A terrorist is merely a rule breaker.

The movie anticipates a related occurrence: the definition of every form of deviant behavior is expansionistic. More and more symptoms and behaviors are brought within the definition. One kind of mental illness (mentioned in an earlier edition of the DSM) is not to admit you have a mental illness. Hence, everyone is mentally ill and must be placed under the control of technology. This is precisely the warning of the movie *Brazil*.

6. Huxley, *Brave New World*, xii.

Chapter 1 _____

Technique Against Culture

Originally published in the *Bulletin of Science, Technology and Society* 15, nos. 2–3 (1995) 73–78.

A re technique and culture incompatible?[1] Jacques Ellul has boldly argued this thesis in many of his works.[2] He did not always state it in exactly these terms; his more usual formulation was that technique destroys meaning. Now the most important element of culture is meaning. If culture, defined as a society's total way of life, both material and symbolic, fails to provide sufficient meaning for its members, is it still a culture?

Why should we be interested in this question? Have not there always been brief periods in the history of a civilization, such as the end of the Middle Ages, when a sense of meaninglessness seemed widespread? Ellul's argument is that today nihilism (meaninglessness) is not a function of the decline of power as in periods of crisis, when a society begins to disintegrate, but is rather a function of the growing power of technique and the political state.[3] In a sense a technological civilization normalizes meaninglessness. If his thesis is correct, then we face a monumental task: the overcoming of technique's natural tendency to destroy common meaning. The impact of this on humans is devastating. Except for the few who have the courage to rise to the challenge, the loss of common meaning produces hopelessness, cynicism, and vague idealism (to cover up the former two tendencies) in the face of life's many problems. And for those, especially the young, who have not managed to repress their hopelessness, self-destructive behavior as with drugs, violence, television, and computer games are forms of escape.

1. Ellul, *Technological Society*, xxv, maintains that "Technique is the totality of methods rationally arrived at and having absolute efficiency (for a given stage of development) in every field of human activity."
2. Ellul, *Technological Bluff*.
3. Ellul, *Betrayal of the West*.

1

The term *meaning* is being used here in the sense of ultimate meaning, the meaning of life, the meaning of history. Meaning implies both significance (value) and orientation (direction).[4] That which is of value has to be realized, to be put into practice, resulting in moral prescriptions and proscriptions. Orientation also involves ethical direction set in time. Certain actions and events, which are prescribed or proscribed, become positive or negative models that took place in the past or will take place in the future. The Holocaust, for instance, is a negative model, to be retained in the collective memory, of what to avoid in the future. The American Revolution, on the other hand, has served as a positive model for a number of developing countries. By contrast Communism articulates an imaginary utopia for the future. The future can be defined as an imaginary one, to be realized for the first time, or as a repetition of some ideal period in the past. Such narratives take the form of myth or history, and provide what Willem Vanderburg calls a "project of existence."[5] Therefore cultural meaning entails some hope for the future.

Related to the question of meaning is that of control or limits. Every culture achieves a certain degree of unity in two related ways: (1) it attributes meaning to certain natural and human activities and relationships so that the negative side of life, such as with suffering and moral evil, can be confronted, and if not overcome, at least resisted, by the positive side as with service, friendship, and love; (2) it places some limitations on the exercise of power—political, economic, technical, personal—thereby preventing a war of all against all, whether at the level of the individual or the group, and allowing societal members to know what to expect of one another. In the following pages, drawing heavily upon Ellul's work, I wish to explore how technique destroys meaning, unleashes power from its former moral controls, and creates false meaning.

Ellul identified three conditions that in conjunction make it virtually impossible for authentic meaning to be created: (1) human relationships become abstract; (2) human activity becomes trivial; and (3) social action becomes ambiguous.[6] All three conditions are a direct result of technique's domination of modern society. The first condition is an expression of technique's mediation of human relationships; the second is a consequence of human powerlessness in the face of technical power; the third condition is a result of the decline of moral communities as technique supplants customs, manners, and morality (shared experiences). After examining how technique has brought about each of these three conditions, we shall turn

4. Ellul, "Life Has Meaning."
5. Vanderburg, *Minds and Cultures.*
6. Ellul, *Ethics of Freedom,* 463–65.

our attention to how false meaning is created as a compensation for the lack of authentic meaning.

One of Ellul's strongest contributions to the study of the technique has been his analysis of nonmaterial technique, those psychological and organizational techniques used to manipulate and control others.[7] Advertising, public relations, therapy, bureaucracy, strategic planning, and the plethora of "how-to" books on achieving "success" in relating to others are but a few examples. Just as technique has been used to exploit the resources of nature in the interest of efficiency, so too has it been necessary to use technique to manage the "human resources" whose natural behavior is always a threat to efficiency. I am not suggesting that these nonmaterial human techniques are always as efficient as their material counterparts; in fact many of them function as forms of magic. But magic can work when people believe in it. This is how many forms of therapy work—the patient believes in the power of the therapist.

What is the main implication of the increasing use of human techniques to mediate human relationships? It makes the relationships abstract and thus impersonal. The objectified nature of technique denies the subjectivity of both user and recipient. As a rational, objective method, technique turns the object of technique into an abstraction. For instance, suppose that as a parent I read the book, *Parental Effectiveness Training*, and decide to rear my children according to this technique. Rather than adjusting my method of child rearing to the individuality of each child and the total context of family relationships, I am now going to use a unitary method, which if applied religiously, will purportedly make each child manageable and thus a standardized product.

The same denial of subjectivity and individuality happens to the user of the technique. Even if it takes substantial effort to learn a technique, once it is learned, the user is relieved of personal responsibility in judgment and choice. In our previous example of the technique of child-rearing, the parent using *Parental Effectiveness Training* gives up learning about his children and exercising moral judgment: he has become an appendage to the technique. It is essential to Ellul's argument about human technique to observe that the subjectivity and individuality of each party in a relationship is threatened by the presence of technique. That is, the subjectivity of the parent is denied in part by not taking into account the subjectivity of his child.

How does turning the user and recipient of technique into abstractions or objects affect the creation of meaning? Humans *create* meaning, spontaneously and interpersonally, because symbolization provides a certain mastery

7. Ellul, *Technological Society.*

over their environment and a peaceful way of relating to one another. Technique denies both spontaneity and genuine intersubjective relationships; hence, it eliminates the preconditions for the creation of meaning. Culture, which revolves around the question of meaning, has been defined as a total way of life. As technique proliferates and is applied to every domain of life, it becomes itself a *total way of life*, an ersatz culture.

The second condition that prevents effective meaning from being created is the trivialization of human activity. The immense power of technique renders human activity ineffective or powerless. Perhaps this is most evident in the world of work. The pursuit of profit through mechanization rendered the worker a specialist, who under the industrial division of labor necessarily made a smaller contribution to the life of the community. The skills of artisans gave way to the technique of the factory, just as the skills of entrepreneurs fell prey to the organizational and psychological techniques of consultants. Technique is not my method; rather it is a rational, objectified method to which I become subject. Technique demands the suppression of human creativity.

Ellul argued that technique is autonomous.[8] What this means is not that humans do not create it, do not modify it, or cannot reject certain forms of it, but that once created the system of relationships among techniques has a life of its own, a certain logic and momentum that we succumb to because of our great belief in technique. That is, we may reject certain techniques, such as chemical or biological weapons, when it becomes obvious that their destructive potential far outweighs any possible gain from using them. Yet the alternatives invariably involve still other techniques. We did not seriously entertain a course of action outside the realm of technique.

As techniques become more powerful and complex, malfunctions are invariably perceived to be due to human error. Humans cannot keep up with the speed, power, and information processing ability of machines, computers, and technological systems. The pervasive but tacit sense of powerlessness today is a reaction to the overwhelming and universal power of technique. This feeling of powerlessness inhibits any attempt to create a common meaning sufficient to regain our control over technique.

But this powerlessness is compounded by our lack of felt need to symbolize technology. Ellul has argued that previously humans' ability to create and live by certain symbols in art, ritual, and religion provided them with the confidence to face the external power of nature and its dangers and eventually gain parity with it.[9] He suggests that this ability to create effective meaning

8. Ellul, *Technological Society*; see also Ellul, *Technological System*.
9. Ellul, "Symbolic Function."

through symbolization has been the paramount factor in our development as a species. We feel no need to symbolize technique because it appears to be our own power. And we do not sense the need to symbolize natural and social problems of disease, moral chaos, and suffering, for we believe technique will solve them.[10] Why symbolize that which is only fleeting?

This brings us to one of the most perplexing problems in understanding technique: our ambivalence toward it. I have argued two apparently contradictory theses: that we feel powerless next to technique and that we perceive it to be our power, not an external power. I think both of these are correct. Technique is an extension of the human and as such is a human power; we *vicariously* share in its successes. Concomitantly, however, the autonomy of technique makes us feel powerless when we measure *our own* personal power against it. Either way, then, in our feelings of powerlessness or in our pride in the power of technique, we did not sense the need to symbolize it and capture it within a web of effective human meaning. And yet symbols abound. As we shall see later, however, these symbols tend to be ephemeral, only capable of providing false meaning.

The third factor that prevents the question of meaning from surfacing is the ambiguity of human action. The decline of a moral community is the chief reason human action becomes ambiguous. Now the erosion of morality implies a loss of ultimate meaning; so the relation between ambiguity in behavior and ambiguity of meaning is reciprocal. What I wish to consider here is how the decline of moral limits and boundaries makes human relationships ambiguous, vague, and dangerous.

From the eighteenth century onward, the prestige of science grew at the expense of religion. Truth eventually was equated with the "objective" findings of science. For a long time people believed that religious and moral values were objective; now they were demoted to the status of the subjective. Ultimate meaning had turned inward. The negative consequences of regarding religious and moral values as subjective were staved off for a while by the remnants of a Judaic-Christian civilization, still strong enough to provide the semblance of a moral consensus. If moral values are subjective, then ultimately morality becomes an expression of political power, either the ability of a group to dominate or to achieve a temporary consensus.

Sensing the impending moral chaos, social Darwinists, pragmatists, and others attempted to ground ethics in scientific findings. If science provides us with truth, then its discoveries about how people behave and develop can become the basis for what ought to be (the ethical). The problem with this, however, is that it is not science so much as technique that becomes the ultimate

10. Dupuy, "Myths of the Informational Society," 9.

criterion of truth.[11] With the eventual subordination of science to technique, truth became in Ellul's words "success relative to reality." This represents the materialization of truth—truth is power.

The increasing centrality of technique eventually led to the decline of the moral community. In the first place technique began to supplant custom, manners, and morality, which traditionally were based on shared experiences and given symbolic meaning. As was previously discussed, nonmaterial human technique becomes a substitute for such shared experiences. In a technological civilization vestiges of custom, manners, and morality prevent many from recognizing the primacy of human technique. Moral custom, for instance, can still be effective as long as the behavior in question does not conflict with the development of technique. Families, for instance, may still practice religion together in a nontechnical way, but increasingly religion in modern societies uses organizational and psychological techniques, such as televangelism and marketing, to reach and control present and future members. Public opinion is solidly behind the expansion and development of technique; it is difficult to resist it. True nonconformity today means resisting the norms of technique rather than traditional moral and legal norms.[12] The upshot of this is that customs, manners, and morality are tolerated but not essential to the order of a technological society.

More is involved, however, than technique merely crowding out shared moral experiences or making them less relevant. There is an inverse relationship between power and values, Ellul notes.[13] Technique is the most powerful, efficient means of action. Modern technique raises power—political, economic, military, psychological—to an unprecedented level so that power itself becomes a value.[14] But how does a growth in power lead to a decline in moral values? The answer revolves around the question of moral limits or boundaries. All moral values place some limits on the exercise of power, collective and individual. Freedom denounces the arbitrary exercise of power by a tyrant. Justice suggests that one is not allowed to take from another whatever one desires. Friendship prohibits the manipulation of one's neighbor. When power becomes too great, it is transformed into a value; moreover, it is a jealous value that permits no other. The discourse about moral values continues (as with the recent discussion of family values) and even increases, but the values cease to be effective values. They are reduced to a kind of ideological distraction that prevents us from seeing reality as

11. Barfield, *Saving the Appearances.*
12. Ellul, *Technological System.*
13. Ellul, *Betrayal of the West.*
14. Ellul, *Technological Bluff,* 94.

it is. The human will to power appears to be stronger than the desire to limit that power, for the growth of power is intoxicating to any group or individual that may benefit from it. The growth of power, then, works to diminish if not destroy the effectiveness of morality.

The decline of a moral community produces human action that is vague, ambiguous, and dangerous.[15] In a community with a shared moral code, human relationships are based on reciprocal expectations; they are safe and predictable most of the time. Not so in a community without a common morality. For here the relationships become more competitive and unpredictable. The other becomes dangerous if only in the sense I cannot trust him anymore. Two nineteenth-century writers, Søren Kierkegaard and Alexis de Tocqueville, observed that this was the first time (outside of a crisis) in which people normally lived in fear of one another.

When human relationships become uncertain, competitive, and even dangerous, the individual protects herself by failing to express her true feelings and even her convictions. Human relationships become an arena in which individuals play a number of roles, feign certain emotions, and manipulate others to their advantage. Much of this is done involuntarily of course. An occasional friendship or family relationship goes beyond this dance of subterfuge to find mutual trust. But for many loneliness is their native land. Capitalism exacerbates the competitiveness that accompanies the decline of a moral community. Karen Horney's insightful *The Neurotic Personality of Our Time* sheds light on the toll competitiveness takes on human relationships. She argues that the distinction between the neurotic and the normal person is becoming blurred. The neurotic suffers from an unconscious attitude that one is "lonely and helpless in an essentially hostile world."[16] More and more of us share this view. She notes that virtually all relationships today, including those of friend and family, are competitive and manipulative. Moreover, because success is a cardinal "value" (success is the moment of power), one only feels good about oneself when one succeeds. This is a "shaky basis for self-esteem."[17] And feeling alone and powerless, we live in fear of each other.

It is not surprising that ambiguous human relationships prevent the creation of meaning. To be effective, meaning has to be created and lived out in concert with others, in an atmosphere of mutual trust. Ellul asserts that if any one of the three conditions that prevent the creation of meaning is

15. van den Berg, *Changing Nature*.

16. Horney, *Neurotic Personality*, 89.

17. Horney, *Neurotic Personality*, 286.

absent, it is still possible albeit difficult to create meaning.[18] Today all three conditions are everywhere in evidence.

Perhaps there are still some who, if they consciously perceived the erosion of cultural meaning, would be willing to struggle to recreate it. False meaning, however, both conceals and compensates for meaninglessness. Hence few penetrate the false meaning to get to the meaninglessness beneath.

What is false meaning? How do we distinguish between authentic meaning and false meaning? Although Ellul was a Christian and had a Christian response to these questions, he explored as well a secular response to them, because the Christian must find some common ground with the non-Christian to make society livable. Authentic meaning is meaning that springs from the lived experiences of people and flows outward so that it is embedded in the hierarchy of authority in that society; false meaning is external, is imposed upon the people, and is inconsistent with their lived experiences. Authentic meaning permits the integration of life's many activities, whereas false meaning is solely a compensation for certain activities that are harmful to cultural and psychological harmony.[19]

We have seen that technique denies the subjectivity of the individual, makes human action trivial, and results in human relationships becoming ambiguous and dangerous. This leads to an extreme discord within the individual and the larger society. What kinds of false meaning does a technological civilization offer its citizens as compensation for this brutal treatment? Consumption. Consumption in every form: goods, services, information, images, and personalities.

As I have argued in *The Culture of Cynicism*, the health and happiness of the human body is the mythological goal of consumption.[20] To consume means to increase. The power of the consumed object becomes my power. Advertising is that institution whose primary cultural function is to provide false meaning by turning every product into a symbol. Products are not merely useful—they must appear to satisfy deep needs, to mollify fears, and to create a new human being. For instance, the sports car symbolizes success; the cologne, sexual prowess; soft drinks, fun and friendship. We consume false meanings every time we consume products and images in this civilization.

These meanings are false in a number of related ways. First, consumer goods and services cannot deliver on their symbolic promises: the sports car

18. Ellul, *Ethics of Freedom*, 463.

19. Sapir, "Culture, Genuine and Spurious."

20. Stivers, *Culture of Cynicism*.

cannot bring you success, at most the trappings of success; the cologne cannot increase your sexual prowess, at best, a minimal and temporary increase in attractiveness.

Second, universal consumption leads to the reification (objectification) of the human being. My needs become objectified and fragmented in consumption. That is, each technological object or product is offered as a material solution to some human need. The number of products is virtually infinite; so too must be my needs. There is no sense here of a human whose needs are integrated, subjective, and finite; but rather of a large warehouse of objectified and atomized needs and desires. The reified human is alienated from himself.

Third, universal consumption leads to the treatment of others as commodities, as consumable objects. David Riesman observed that the consumption of other people's personalities had become an American preoccupation.[21] To consume here means to treat the other's personality as an object that exists for my enjoyment and entertainment.

False meaning appears conspiratorial at its source, but is experienced as random in its consequences. False meaning is not the result of shared experiences that we make meaningful, but is inflicted on us in acts of manipulation that isolate us psychologically. The mass media reduce us to individual consumers of random information and spectators of random images. In turn the proliferation of such information and images makes life and human relationships appear random. Appearances to the contrary, advertising, which is the chief dispenser of false meaning, is not guided by a conscious conspiracy to supplant genuine meaning with false meaning. Rather it evolves as part of the technological system without decisive moral and political control. As genuine meaning declines, life is perceived, at least tacitly, to be like life in a Thomas Pynchon novel—both arbitrary and random.

Finally, Ellul notes that false meaning does not effectively compensate for meaninglessness, but actually reinforces it; for false meaning is transitory and illusory.[22] False meaning constantly lets us down; our hopes are always being dashed. And the more meaninglessness deepens in us, the more we lose the will to resist and overcome it. False meaning gushes from every pore of our technological civilization. All this occurs because we tacitly regard technique as sacred, that very technique which destroys genuine meaning and for it substitutes false meaning. For those of us who still care about culture, the first step is to desacralize technique, to destroy its credibility as our savior. In doing so we would be honoring the memory of Jacques Ellul.

21. Riesman, *Lonely Crowd*.
22. Ellul, *Ethics of Freedom*, 469.

Chapter 2 _____

The Media Creates Us in Its Image

Originally published in the *Bulletin of Science, Technology and Society* 32, no. 3 (2012) 203–12.

I n Mary Shelley's *Frankenstein*, Dr. Frankenstein created a human-like monster (who also came to be known as Frankenstein). We too have created a monster—the media—but a monster who in turn creates us in its image. Our monster does not terrorize us, but happily and gently makes us conform to its dictates and assumptions about what it means to be human.

Søren Kierkegaard understood the essential points about public opinion and the media in the mid-nineteenth century when the newspaper and pamphlet were the paramount media.[1] Despite additional media such as television, the Internet, and social media, my thesis is that nothing has changed qualitatively. The media today gather an enormous amount of information, process it, and transmit it with great speed, but these changes are only quantitative. Kierkegaard's insights remain valid.

In the second half of the twentieth century, Jacques Ellul extended Kierkegaard's analysis with his study of propaganda and the visual images in the media.[2] Kierkegaard had only referred to propaganda (advertising and publicity) in passing.[3] In Ellul's view, propaganda is the primary means by which the technological system controls us.[4]

At the same time Ellul completed his definitive study of propaganda, George Trow wrote a most suggestive book about how television psychologically captures us in its redefinition of reality. Trow's analysis extends beyond television to other media.[5]

1. Kierkegaard, *Present Age*.
2. Ellul, *Propaganda*; Ellul, *Humiliation of the Word*.
3. Kierkegaard, *Present Age*.
4. Ellul, *Propaganda*.
5. Trow, *No Context*.

Wolfgang Schivelbusch supplied a missing concept in understanding technology's psychological impact on us.[6] The "stimulus shield," a term he borrowed from Freud, refers to our tendency to internalize and thus normalize technological stimuli. Technology alters our sense of time and space, and we adapt to these changes. For example, after railway travel became customary, people had trouble returning to slower travel by horse: they had internalized the speed of the train.

Kierkegaard studied the public and its opinion, Ellul, propaganda and mass society, Trow, television, and Schivelbusch, the way we internalize technology. Public opinion, propaganda, and the media are mutually dependent on each other and thus form a cultural configuration. The concept of the stimulus shield indicates that the media's impact on us may be much greater than we realize. Indeed, the media has transformed the reality of what it means to be human even if some of us cling to an outdated ideal.

The Ideal Human in Western Civilization

Every civilization creates a social character type, that is, sets the cultural parameters within which individual emotional and intellectual development occurs. Western civilization includes Greek, Roman, and Christian influences. The Greek and Roman part of Western civilization on one hand and Christian aspect on the other hand contradict each other. But a synthesis of Christian and Greek thought occurred in the Middle Ages, and slightly later in the Renaissance, a synthesis of Roman and Christian thought.[7] Western civilization eventually reached its apogee in the Enlightenment.

The resulting syncretism placed an emphasis on reason, individuality, and freedom. This creation took centuries with, however, reversals and periods of stagnation. Norbert Elias refers to it as the "civilizing process," but gives it a different interpretation.[8] Elias attributes the gradual internalization of manners and morality and resultant self-control to the growth and centralization of power in the political state. For Ellul the gradual creation of the values of reason, individuality, and freedom owe more to Greek, Roman, and Christian cultural influences than to the centralization of the military function in the state.[9]

Reason was not a specialized technical reason but a normative reason that featured self-reflection. Reason was not an end in itself but served moral

6. Schivelbusch, *Railway Journey*.
7. Ellul, *Betrayal of the West*.
8. Elias, *History of Manners*.
9. Ellul, *Betrayal of the West*.

and communal purposes as well as individual ends. Reason became aberrant when it denied the emotions a part in deliberation. Reason meant that the individual thought about the meaning of the symbolic dimensions of culture, and that consequently society created rational social institutions.

Reason both facilitated and was made possible by a growing sense of individuality, which necessarily implied a conflict between the individual and society. Hence, the individual could resist excessive power, whether political or religious, as his right. The individual was now more than a biological individual but a moral person who was both responsible and free. Freedom was an expression of individuality, but only when it entailed both responsibility and self-control. The idea of civil disobedience is perhaps the highest form of freedom: one recognizes society's right to establish laws, but concurrently the right of the individual to violate an unjust law if she is willing to suffer the punishment as a protest.

The coming together of reason, freedom, and individuality produced a consistent, coherent human being with moral core to the self. Moral self-control resulted in an individual whose consistency in the living out of his beliefs made him coherent to others. Consistency and coherence are essential to trust.

Clearly, there were dangers in the way this character type was put into practice. Reason could deny emotion, and moral and rational rigidity could restrict freedom and a love of others. But at its best this character type helped to create an individual who was both responsible and free, capable of making distinctions that avoided narrowness of mind and self-centeredness.

In the second half of the twentieth century, attacks on this Western character ideal came from a variety of sources. Structuralism was an academic movement in the humanities and social sciences that denied the autonomy of the individual and the effectiveness of reason. In its attack on Enlightenment thought, structuralism's battle cry was "death to the subject." Unconscious cultural forces, it was claimed, obliterated the freedom and reason of the individual. No sooner had structuralism begun to wane then postmodernism continued the assault on Enlightenment assumptions about reason, morality, and language.

Postmodernism rejects the ability of individual reason to interpret discourse validly. The inherent ambiguity of meaning in natural language, postmodernism argues, makes a perfect interpretation impossible; it even rejects the idea of better and worse interpretations. Postmodernism represents an attack on language and the concept of truth.

Culture necessitates a hierarchy of aesthetical and ethical values. Cultural authority is another name for hierarchy. Postmodernism rejects cultural authority in the name of cultural equality. At its extreme postmodernism regards all texts, all interpretations, all art forms, all moralities, all meanings as equal. A plethora of interpretations compete for attention; none can dominate for long. Literature, art, and language reflect this fragmentation and movement toward nihilism. Cultural equality (the absence of authority) is equated with freedom.

The animus of postmodernism is directed against the ideal of truth. When reason and language do not permit us to establish truth, then truth becomes political, that is, it is imposed on us by those with power.[10]

If structuralism and postmodernism were intellectual movements in the humanities and social sciences, posthumanism is a product of science and technology. Posthumanism is indebted to cybernetics, cognitive science, information theory, and artificial intelligence. The first and lasting comparison is between the human and the computer. This comparison does not restrict itself to the problem of intelligence. Perhaps, some argue, human emotions can be mechanized as well. The wager, posthumanists make, is that by treating nature as a vast computing machine, science and technology can make rapid progress. Rather than stop with this comparison, however, inadvertently the proponents turned computerized reality into a means for model building. The most extreme advocates of cybernetics were caught in an enormous tautology: nature is compared to a computing machine and we mistake reality for a scientific model. Scientific models of computer software have become reality.[11]

N. Katherine Hayles has identified the key component of posthumanism: (a) Information is more important than its embodiment, (b) human consciousness as the foundation of human identity is a small and minor part of human evolution, (c) the body is a prosthesis and replacing it with still other prostheses is part of progress, and (d) the human is an intelligent machine interchangeable with other intelligent machines.[12]

Biological determinism represents another attempt to minimize the role of culture in creating a rational and free individual. Geneticist Richard Lewontin describes the ideology of biological determinism: (a) Variations in human capabilities are due to innate differences, (b) the latter are biologically inherited, and (c) that the resultant inequality in society is a function of biological differences.[13] Genetic determinism is a form of biological determinism. Not only do we not know in a strict sense what a gene is, but genetic determinism omits the "temporal sequence of external environments"

10. Stivers, *Illusion of Freedom and Equality.*
11. Dupuy, *Cognitive Science.*
12. Hayles, *How We Became Posthuman.*
13. Lewontin, *Biology as Ideology.*

through which the gene passes in its life" and "random events of molecular interactions within individual cells."[14] This is why, Lewontin argues, "even with the complete DNA sequence of an organism and unlimited computational power, we could not compute the organism, because the organism does not compute itself from its genes."[15] Truly interactive factors cannot be controlled and measured. Reducing nature to a code that a computer can define and measure is a myth.

What has prompted these attacks on the Western definition of the human, especially the Enlightenment ideal? There are two interrelated reasons for the attempt to reduce the human to a machine or an organism: nihilism and the identification of truth with technology.[16]

The triumph of science and technology is coeval with the experience of nihilism. All cultures are religious at their basis, that is, they are organized into a hierarchy of values, of which the sacred is the highest value. Moreover, religion provides a worldview, a view of reality perceived to be both objective and true, and an ethos, a definition of the meaning of life and the meaning of history. Religion furthermore shapes common sense so that religious beliefs do not appear extraordinary.[17] In the Middle Ages, visual art revealed a reality peopled with spiritual beings, angels and demons, and saints and sinners. Reality was more than physical reality; it was infused with meaning.

In the nineteenth century science began to overtake religion; the former was thought to be objective and the arbiter of truth. Science escapes the laboratory and becomes a worldview, that of "radical immanentism." Radical immanentism assumes we live in a self-contained material world with no transcendent meaning. At its deepest level, Gabriel Vahanian maintains, Western civilization became atheistic.[18] For most people this view needs to remain unconscious, as Nietzsche understood. We overlay it with religion that becomes subjective. Religion was increasingly seen as something personal—for my family and me—and thus could be a comfort for the bleak material reality that science offered.

When religion and values become subjective, nihilism ensues. The meaning of life and the meaning of history have to be common and understood as objective to be effective. Existential nihilism does not so much refer to the absence of values or meaninglessness as to the fragmentation

14. Lewontin, *Biology as Ideology*, 23.
15. Lewontin, *Triple Helix*, 17–18.
16. Barfield, *Saving the Appearances*.
17. Geertz, *Interpretation of Cultures*.
18. Vahanian, *Death of God*.

of values. Either way values lose their efficacy. Science cannot provide lasting values because the meaning of science resides in technology. Modern technology is concerned with the most powerful or efficient means of action. Technology attempts to turn power into a value, but cannot do so. For as power increases, the efficacy of values decreases. That is, all ethical values place some limitation on power. Justice, for example, means that I cannot take what is not mine. Consequently, making power a "value" creates cynicism, the view that all life is a struggle for power. The motto is, manipulate or be manipulated.[19]

Now cynicism and nihilism result in hopelessness, which in turn creates despair. This despair assumes many forms: apathy, self-destructive behavior, and a hatred of culture.[20] Indirectly, then, structuralism, postmodernism, posthumanism, genetic determinism, genetic engineering, and the creation of cyborgs (merger of human and machine) represent an existential nihilism that places no limitations on the destruction of culture (the ideals of individuality, reason, and freedom) and experimentations on humans. As such they are the ideological expression of the triumph of the technological system over a living culture.

For the ordinary person, however, structuralism, postmodernism, posthumanism, and genetic determinism are academic controversies and of little interest. What is frightening, however, are sensationalized issues such as cloning, genetic engineering, and biotechnology. The technological manipulation of human beings, not science, makes us uneasy. The merger of human and computer and the apparent inferiority of humans portends the end of humanity. Jacques Ellul once noted that fantastic, even science fiction-like technological threats, distract us from the way technology is already changing us.[21] And the most important of these is the media.

The Media in the Context of Public Opinion, Propaganda, and the Mass Society

We need to examine the relationships among public opinion, the mass society, propaganda, and the media if only in a preliminary way. Propaganda has existed well before the emergence of the technological society. Modern propaganda, however, requires a mass society, in which public opinion dominates private opinion, and in which the mass media of communication both creates and follows public opinion. In other words, we are studying

19. Stivers, *Culture of Cynicism.*
20. Ellul, *Time of Abandonment.*
21. Ellul, *Technological Society.*

information and communication in the technological society. The formation of mass societies was necessary for the emergence of a technological society. A technological society is held together by public opinion that is created especially by sociological propaganda (advertising and public relations) and transmitted by the mass media.

Mass Society

Ellul identifies a mass society as one of the necessary prerequisites for the emergence of a technological society.[22] A mass society is one that is simultaneously individualistic and collectivistic. Power becomes abstract and concentrated in the political state, large organizations, the media, and technology. Concurrently there is a decline of cultural authority that resides in the community and extended family. Because the individual in a traditional society was responsible to others in family and community, she was not an isolated individual. She could depend on others for assistance. With the decline of cultural authority our relationships with others become ambiguous.[23] A common morality is the basis for trust, and when morality erodes, we live in fear of others, not so much physical fear, as fear of what they think about us. We cannot count on others anymore with any certitude. Consequently, Alexis de Tocqueville argued, in our weakness and fear we turn to government and public opinion to protect us. Individualism ultimately reinforces collectivism.[24]

A mass society is segmented into special interest groups on one hand and peer groups on the other. The peer group represents an intermediate stage between the personal relations of family and community and the more impersonal relations within special interest groups. Certainly, there have always been groups based on age and sex. In traditional societies, such groups were integrated into the community. The peer group becomes more or less autonomous, however, in a mass society. Consequently internal relations become competitive in respect to popularity and dominance even though the peer group provides partial protection against the fear and loneliness of the individual in a mass society.

Even though a common morality is on the wane in a mass society, similar mythical beliefs and an interest in technology help create a fragile unity. Public opinion is the critical unifying force, however. The peer group's most important function is to contextualize public opinion for the individual.

22. Ellul, *Technological Society*.
23. van den Berg, *Divided Existence*.
24. Tocqueville, *Democracy in America*.

Public Opinion

Before examining public opinion, we will first discuss what Kierkegaard meant by the public, concentrating on its social psychology.[25] The public is formed by individuals being detached observers of others and by the mass media. This condition of being an onlooker to life involves a complex of intellectual and psychological processes: reflection, envy, and moral resentment.

In a period of moral dissolution, relationships between individuals become ambiguous in that the aesthetical interests of each party are in conflict and yet remain partially concealed both to self and other. It is not polite to admit one's selfish motives. This leaves the individual in a state of "reflective tension." Rather than an inward relationship regulated and given meaning by moral qualities such as trust and self-sacrifice, the relationship has become external. In Kierkegaard's formulation, passion can only be expressed by those whose relationship to others is inward, that is, ethically qualitative. With the decline of inwardness and passion, one's relationship to the other becomes concurrently that of aesthetic possibility and ethical indifference. As such, the relationship must necessarily become abstract, the object of theoretical reasoning (reflection). Life in a "reflective age" assumes the characteristics of a game in which one plots one's moves in advance in order to maximize one's chances for success.

Reflection itself, Kierkegaard maintains, is not the problem, but rather reflection that does not result in action, reflection that becomes an escape from action.[26] When reflection is not accompanied by inward moral commitment, envy ensues. The powers of reflection are at the disposal of selfish desires. One envies others, their possessions and accomplishments. "Selfish envy" actually deters the individual from decisive action; instead time is spent ruminating about one's possibilities or what it would be like to be someone else. At this point, Kierkegaard draws a startling conclusion about the modern reflective and passionless age: "envy is the negative unifying principle."[27] This envy has both individual and collective manifestations, for the envy within the individual has its counterpart in the envious attitudes of others. As we shall shortly see, the chief way in which this collective envy is expressed is through public opinion.

When the envy that is present in reflection as aesthetic possibility is not punctuated by decision and action, it spills over into moral resentment.

---

25. Kierkegaard, *Present Age*.

26. Kierkegaard, *Present Age*.

27. Kierkegaard, *Present Age*, 47.

Assuredly resentment is present in every age but what makes it different in the modern era is "leveling." By this Kierkegaard means the attitude that no one is better than I am.[28] Because of a lack of moral character the individual denigrates and even ridicules those who have distinguished themselves and have moral authority. It is not enough to admire and envy the other; one must tear him down. Leveling to be effective must be done in concert with others; it is essentially a collective phenomenon. Whereas it was once the province of a social class or occupation, in a reflective and passionless age it is accomplished by the public. The public is an abstraction in that the members do not interact with one another; therefore, the public's opinion must be expressed through the mass media. Because moral resentment takes the form of leveling in a time of passionless reflection, and because leveling is expressed through public opinion, public opinion both creates and is an expression of the "negative unity of the negative reciprocity of all individuals."[29] Public opinion has a variety of characteristics. First, it expresses desire and fear. Second, it is based on secondhand knowledge rather than on experience or readily intelligible facts. Public opinion is a necessary component of a mass society; it is a substitute for interpersonal knowledge that arises from practical, everyday life in the context of intense familial and friendship relations. The mass media provide us with ersatz experiences upon which opinion rests. If, as Daniel Boorstin notes, personal knowledge is phenomenal, public opinion is epiphenomenal; it is concerned with what most of us know little about through experience or serious thought.[30] Therefore, public opinion is secondary both in regard to its source—the mass media—and its type of knowledge—epiphenomenal.

More obvious than public opinion's artificial character is its fragmentation. There is neither coherence nor continuity to the totality of opinions. A sure sign of fragmentation is the inconsistent if not contradictory quality of many opinions. For example, American opinion supports preserving the environment and economic growth simultaneously. Opinions are always out of context, for issues are presented to the public consecutively as self-contained entities; they resemble facts that appear to be autonomous. The fragmentation of opinion is a reflection of the fragmentation of information conveyed through the media. Infatuated with the new and the sensational, the media unintentionally destroy the memory of the past.

In part as a consequence of fragmentation, public opinion is based on simplified issues. Complex and profound problems can only be resolved, and

28. Kierkegaard, *Present Age.*
29. Kierkegaard, *Present Age,* 52.
30. Boorstin, *Democracy and Its Discontents.*

even then only rarely, through long and arduous discussion. Because opinion forms (or is pressured to form) on an enormous number of issues, many of which are outside of most people's personal expertise, the media must simplify the issue. Only then can public opinion emerge. Criticism of American politicians, for example, sometimes centers on their tendency to simplify issues, when the problem is actually much more a result of the mass media that can only operate profitably through simplification, and the public, which, no matter how well educated, due to technical specialization, is ignorant about most issues. Inefficient bureaucracy, for instance, comes under attack from the political right in its assault on bloated government, when the evidence is rather strong that bureaucracy under certain circumstances is highly efficient and that big business is itself highly bureaucratized.

A prevalent form that simplified opinion assumes is that of the stereotype. Stereotypes need not be negative, indeed, they are just as often positive. In a time when we are expected to have an opinion on everything and everybody, generalizations based on limited experience and a handful of cases abound. Stereotypes in a mass society permit people with limited time to place people and events into a manageable set of categories.

Public opinion forms itself around issues that are unrelated to reality. To a certain extent this can be deduced from its fragmentary and secondary nature. Public opinion is knowledge taken out of historical and cultural context. Just as important is the symbolic dimension of these issues. Public opinion is not created out of common interpersonal experiences; hence, those whose views form public opinion must be supplied with a common denominator—mythical beliefs. Propaganda is the context of such mythical ideals.[31]

Propaganda

As we have seen, modern propaganda depends on strong public opinion and a mass society for its existence, and of course the mass media. Modern propaganda is a form of technology. It is a rationally constructed method that aims at efficiency in the manipulation and control of individuals in a mass society. Propaganda's success depends on its ability to work on our emotions in a symbolic and unconscious way: propaganda manipulates us before it attempts to persuade us rationally.

As such propaganda is bound to the universe of power and efficiency. Ellul observes that modern propaganda is radically different from traditional propaganda. No longer does propaganda follow ideology; now

31. Ellul, *Propaganda.*

ideology is just a screen for propaganda. That is, ideology is used in the service of manipulation to bring about action. Then too propaganda follows events. First the event, for example, the invasion of Iraq, then the propaganda campaign in earnest. Propaganda is completely dominated by the will to power: it can rationalize a set of actions either before or after they occur. Propaganda, like all techniques, is autonomous in regard to morality. For propaganda to be effective it must be continuous, widespread, and in step with public opinion and the unconscious beliefs of a society. Propaganda does not automatically succeed.

One of Ellul's greatest contributions to the study of propaganda is the distinction between political propaganda and sociological propaganda.[32] The former of course is propaganda employed by the government, a party, or a movement. The latter is less well known. Sociological propaganda refers to advertising and public relations, and the rest of the content of the media. It also refers less obviously to the human sciences, the socializing aspects of education, and information. Sociological propaganda is more diffuse than political propaganda and is not the work of an elite group of propagandists. It is replete in all social institutions and contains a "general conception of society, a way of life."[33] He offers as an example the American Way of Life. Will Herberg has described the American Way of Life in *Protestant, Catholic, Jew* as the actual religion of Americans.[34] It entails a belief in American free enterprise and democracy and the search for health, happiness, and success. In short, the American Way of Life is a belief in everything American.

Sociological propaganda now forms culture even if a fragmented one. Daniel Boorstin noted this as well when referring to advertising as the American culture.[35] Ellul described how propaganda is less dependent on ideology, instead employing myth, symbols, and stereotypes.[36] So if sociological propaganda is our culture there must be a unifying myth. Ideology separates, myth unites. Communism, socialism, and capitalism are ideologies in conflict, but they all subscribe to a common myth.

I have termed the common myth the myth of technological utopianism. All societies have myths of origin, the creation of the universe, and the founding of a society. These myths center on what people unconsciously regard as sacred. Today people tacitly regard technology as

32. Ellul, *Propaganda*.

33. Ellul, *Propaganda*, 65.

34. Herberg, *Protestant, Catholic, Jew*.

35. Boorstin, *Democracy and Its Discontents*.

36. Ellul, *Critique of the New Commonplaces*.

sacred.[37] The myth of technological utopianism is the foundation of sociological propaganda in modern technological societies. It is especially strong in America because our identity has been shaped by technology since the nineteenth century.[38]

I will be drawing the following remarks from my *Technology as Magic*.[39] The narrative of the myth of technological utopianism is straightforward. Science and especially technology are leading us to a utopia of maximum production and consumption. Technology insures our collective survival and success in allowing us more efficient control of life and providing solutions to all our problems. This promised land is likewise a world of total consumption. In it people have perfect health, are beautiful, eternally youthful, free to do whatever is pleasurable, and thus completely happy. The myth of technological utopianism is promulgated through the liturgy of advertising.

The myth contains four main symbols (all myths are composed of symbols, which give meaning to the narrative): success, survival, happiness, and health. The former two symbols refer to the collective dimensions of a technological society; the latter two refer to individual consumerism.

If the world of advertising is truly a mythological world, then it exists outside of the dialectic of truth and falsehood. For the world advertising creates is not actual but only possible. As with all mythologized rituals, advertising can withstand the negative test of reality for there is always a next time: the possibility of perfection and total fulfillment in the newest commodity. Myth likewise works to overcome contradictions that we experience in the everyday world. The technological system creates cultural meaninglessness and intense psychological stress. This system does not rest easy upon human society. As well, technological growth threatens the physical environment. Consequently, some of the technology has to be directed toward helping individuals adjust to the system and toward repairing the damage to the environment. But this can only be done if adjustment is brought within the symbol of happiness, and repair (survival) is contained within the symbol of success. As part of the myth of technological utopianism, all four symbols are interrelated; moreover, each one implies the rest. The value of success was gradually transformed from an individual to a collective phenomenon. Success became by the late nineteenth century the success of the organization. The equation of the American nation with technology meant that national progress was insured by technological growth.

37. Ellul, *New Demons*.
38. Stivers, *Culture of Cynicism*.
39. Stivers, *Technology as Magic*.

The value of success is related to that of survival: both are expressions of collective power turned into a value. Survival is minimalist success. If success today is most epitomized in technological growth, then survival is related to the destructive aspects of that same growth. The value of survival grows increasingly important as we become acutely aware of problems such as pollution, overpopulation, and potential nuclear catastrophe that require repair. Success (as technological growth) stands in a contradictory relationship to survival (as technological repair).

As success is collectivized in technology, it is redefined for the individual in terms of well-being (happiness and health). During the twentieth century, happiness and health have each come to possess two distinct meanings. Happiness refers to the consumption of goods and services and to adjustment to one's circumstances in life; health refers to the perfection of the body through consumer goods and services, for example, vitamins and organized exercise, and to adjustment, defined as emotional or mental health. (The most prevalent criterion of mental health for much of this century has been adjustment.) Therefore, happiness and health have a common meaning in adjustment; concurrently happiness as consumption and health as the perfection of the body share the common meaning of physical well-being. Happiness and health (taken together) have two overall meanings: physical well-being and emotional well-being (adjustment). The overall meanings, moreover, are related. First, a consumption-oriented lifestyle is a major part of adjustment to the technological system. It is our compensation for the diminution of moral responsibility and individual freedom. Second, physical well-being and emotional well-being are increasingly perceived to be interdependent. But just as success as technological growth can threaten survival, happiness as consumption can impair one's physical or emotional health.

The mythological symbols are arranged in a hierarchy: Success and happiness are higher in value than survival and health (as repair and adjustment). The latter two symbols indirectly reflect technology's inability to carry through on its utopian promise. Between the former two symbols, happiness is higher, at least from the individual's viewpoint. That is, technology can grow indefinitely, but if it does not lead to individual happiness, what good is it? Keep in mind, however, that if technology because of its interconnections has become a system and thus autonomous, it is not directed toward individual happiness but only its own continued growth and survival. In the mythological world the various symbols are compatible; in the real world their realizations are at odds.

The myth of technological utopianism does not occur only in advertising; it is also contained in media programs. Advertising is pervasive. Take television, for example. The television programs are longer versions of ads so that television programs are ads for advertising. If the symbols success, survival, happiness, and health are present in advertising, the programs feature longer narratives that express these symbols.

If you think of television programs as forming a complex, then each program does not have to express the four symbols; rather the totality of programs does. There are four genres of program, each of which provides a mythological symbol: sports is about success, the news about survival, children's shows about happiness, and soap operas about health. The other genres contain a combination of symbols.

Does technological utopianism suffice as a source of meaning for modern societies? If so, does this not contradict the earlier discussion of nihilism? Power in itself cannot provide meaning. The latter arises in the symbolic meaning given to and the moral control exercised over power. Modern symbolism is dominated by visual images and is meaningless. Visual symbolism creates no meaning, for it is the result of literal or material associations. In the famous Coca-Cola ad in which children from all over the world are holding hands, singing about peace and harmony, and finally drinking Coke, peace and harmony equal holding hands, and holding hands equals Coca-Cola. Life consists of material relationships to people and to products. But the "meaning" of life resides in that which is powerful—technological objects such as Coca-Cola. In a traditional symbol, such as love is a rose, the meaning arises out of the tension between literal and figurative meanings. Love is *not* literally a rose. In visual advertising, *peace and harmony* are literally Coca-Cola.

In the myth of technological utopianism success and survival refer to the power of technology as production, happiness and health to the power of individual consumption. Technological utopianism creates no meaning; it throws us back on the material power of technology.

The myth of technological utopianism provides an anchor for sociological propaganda and thus forms the basis of culture. Sociological propaganda is the true rationale for the mass media; the two are inseparably linked. Sociological propaganda is the content of the media.

The Media

The media makes a contribution apart from its content of propaganda to the manipulation of the masses: its form.[40] In the mid-nineteenth century, Kierkegaard noted that the amount of information (meaningless noise)

40. Ellul, *Propaganda*.

and the speed by which it was communicated had considerably increased.[41] What would he think about the technological milieu today?

Social media, mass media, cell phones, email, fax machines, the Internet, conference calls, and so forth create a technological environment of continuous communication. The average child spends well over six hours each day using media of all kinds.[42] Sherry Turkle's research indicates that young people especially feel an insatiable need to stay connected to one another.[43] My students tell me that they need to have media on in the background when they study or work in their apartments. The media not only keep us connected but serve as white noise to distract us from our stressful lives.

The visual image is dominant in the media, not just on television and in video games but also in other media that depend more on language. All natural languages are ambiguous; words do not have a single meaning. The context of discourse reduces the ambiguity. The meaning of words is thus in their usage, which is set in a context: the sentence, paragraph, chapter, book, or the dialogue. The ambiguity of language allows us to create new meanings, as with a fresh metaphor. Today, however, language is going in opposite directions. Technical terms have one precise but operationalized meaning; other words have become vague. The vagueness of words cannot be overcome by context because the meanings of such words have been divorced from what they refer to in reality. For instance, if the word *democracy* can be used to refer to any dictatorship as long as it is friendly to us, then the word is being used for propagandistic purposes. Political and sociological propaganda are using words to manipulate the public; hence, words must become vague. Visual images objectify the meaning of vague words. Therefore, the meaning of love becomes the visual image of a hug or a kiss.[44] Visual images in the media have subjugated language.

Visual images first of all hit us on an emotional level. Emotional experiences are largely aesthetical, and as such orient us to the pleasure or pain of the moment. Aesthetical action refers to living in the moment of pleasure, to how the world and people affect our senses. Is it beautiful or ugly, interesting or boring? These are aesthetical questions. The media aestheticize reality. The point of the media, Neil Postman observed, is amusement.[45] Søren Kierkegaard claimed that to develop a completely aesthetical

41. Kierkegaard, *Present Age*.
42. Gitlin, *Media Unlimited*.
43. Turkle, *Alone Together*.
44. Stivers, *Culture of Cynicism*.
45. Postman, *Amusing Ourselves to Death*.

lifestyle is to be ethically indifferent to others (pleasure is selfish) and to remain a child.[46] Children have to be socialized to get beyond pleasure. The media is a pleasure machine.

The images in the media are fragmentary. Each image is a totality in itself. Visual images by themselves destroy narrative structure; only a text can provide it. Television moves us from program to commercials and back to program, from program to program, with no regret. The result is a fragmentary, random view of life. Only narrative and morality can provide a cultural unity. An aesthetical approach to life is moment to moment, with each moment an end in itself. Fragmentary images are the tool of propaganda, which attempts to make the masses respond to images, symbols, and stereotypes immediately and emotionally.

The media creates an overload of information, which becomes disinformation and leaves us with a "broken vision of the world." We are unable to integrate the surfeit of information we receive and as a result lose the ability to distinguish between relevant and irrelevant information. All information tends to become equal; nihilism ensues. Because most of this information is about the present, we lose the capacity to situate it in a historical context. Consequently, the information becomes more fearful. Finally, an excess of information transforms us into consumers of information, constantly searching for the newest, most sensational, and most relevant information.[47]

The enormous quantity of information is matched by the speed at which it is processed and transmitted. The tempo of media commercials, programs, video games, and films has accelerated. Commercials contain the most hyperactivated imagery. There are an average of ten to fifteen image changes per thirty-second commercial and seven to ten per sixty-second segment during a typical television program. The tempo of films has likewise increased. Many people complain about slow-moving films that feature dialogue and character development. The faster and more frenzied the action, the better. Events and time are accelerated, condensed, and simplified.[48]

The computer helps create a synchronous society by reducing time to ever smaller units. The computer produces the experience of instantaneous time as it processes increasingly enormous amounts of information in nanoseconds. As Ellul observes, "Real time, in which the computer now functions, is a time looped in advance and made instantaneous."[49] With the assistance of the computer, the media "programs" time, that is, it subordinates

46. Kierkegaard, *Present Age*.
47. Ellul, *Technological Bluff*.
48. Stivers, *Shades of Loneliness*.
49. Ellul, *Technological Bluff*, 95.

time to its own functioning rather than to human existence. Therefore it reduces the past and the future to the present. But because the present is shrinking to the nanosecond, technology appears to exist outside of time. Qualitative time, by contrast, exists in the narrative form of past, present, and future. Meaning and values only exist in qualitative time. "Speed is the form of ecstasy the technological revolution has bestowed on man," wrote Milan Kundera.[50] More precisely, constant and rapid stimulation is the form of ecstasy we welcome and even demand.

The media exist outside traditional cultural and historical contexts; they create their own contexts and their own reality. George Trow has identified the contexts of no context: problem and solution, false love, and the celebrity.[51] Problem and solution is the domain of the expert. When cultural authority declines, the expert takes its place. The expert is a technician who bestows specialized knowledge on those afflicted with a problem. Trow says that the authority of the expert lies in the problem. The media features all kinds of experts, such as Dr. Phil, especially on talk shows.

Trow refers to television as the "cold child."[52] The cold child exudes a pseudo-cheerfulness. Media personalities appear to be interested in us, even love us, but it is only pseudo-intimacy. The emotions displayed on television are an affectation—they are manipulative. This is why the celebrity is the key in integrating the individual into a mass society. Only the celebrity has a complete life, that is, a personal existence and a mass existence. Because reality is now in the media, we vicariously experience all the triumphs and failures of celebrities.[53] Our lives become infinitely more interesting as a result. Thus, the celebrity is a pseudo-cure for loneliness. Moreover, the celebrity is an expert in living. She is a model to emulate. Our impersonal relationship to celebrities indicates that we accept shallow human relationships as normal.

Television, Trow argues, has a short attention span, like a child. This is another way of understanding the fragmentary nature of the dramatized information in the media. The cold child is immature in its preoccupation with fun. Television and all the media doom us to a life sentence of nonstop and rapid amusement.

The discussion so far treats the media as a unitary phenomenon. Do the widely popular social media make relationships more personal than those we have with celebrities? Sherry Turkle's recent study of the social

---

50. Kundera, *Slowness*, 2.

51. Trow, *No Context*.

52. Trow, *No Context*.

53. Ellul, *Humiliation of the Word*.

media indicates how little difference there is between the mass media and social media.[54] The social media speed up our relationships, and as with Facebook and Twitter, increase their number. Turkle notes that many young people fear intimacy and have no clear conception of privacy.[55] Some even state that they prefer texting to phone conversations because the former do not commit you to the other person. The social media can create group feelings, whereby one has to validate one's feelings with others. The social media creates a high degree of conformity. Despite this the social media leaves us lonely, as does extended use of the computer. As Turkle wisely observes, "Loneliness is failed solitude."[56] Intimacy requires time alone to understand one's moral and emotional commitments and reaffirm their value. A lack of privacy and the experience of "group feelings" sentence one to shallow and ephemeral relationships. The mass media and social media both produce conformity and loneliness and leave us with shallow relationships. The relationships in the social media are less anonymous most of the time, but this is hardly a comfort. Shallow anxious relationships are only slightly better than anonymous ones.

The Creation

What kind of person does the media create? Before answering this question we need to return to the concept of the stimulus shield. The concept refers to the need for and ability to internalize and normalize technological stimuli that initially create fear and anxiety. We adapt to new technologies that alter our sense of time and space. We reduce our anxiety by becoming like the very technology that initially frightened us.[57]

Studies indicate that employees who frequently use computers sometimes display irritation with people and machines that do not perform quickly enough. One researcher concluded that the "worker internalizes the rapid, instant access mode of computer operations."[58] The media in general speed up our perceptual and nervous systems so that we seem "bored" with a slower pace of life and time alone to read, think, and be introspective.[59]

Nicholas Carr cites research that indicates that the mental functions that appear to be diminishing because of a lack of use are those that support "calm,

54. Turkle, *Alone Together*.
55. Turkle, *Alone Together*.
56. Turkle, *Alone Together*, 288.
57. Schivelbusch, *Railway Journey*.
58. Brod, *Technostress*.
59. Mander, *Absence of the Sacred*.

linear thought."[60] They are giving way to those functions that help us speedily locate, categorize, and assess disparate bits of information in a variety of forms that let us maintain our mental bearings while being bombarded by stimuli. These functions are, not coincidentally, very similar to the ones performed by computers, which are programmed for the high-speed transfer of data in and out of memory. Once again, we seem to be taking on the characteristics of a popular new intellectual technology.[61]

In the past few years, there has been an increase in research on distraction and multitasking.[62] We become distracted when our attention to a specific task begins to wane. The quantity and speed of information transmitted in the media is a cause of distraction. Multitasking allows us to work while distracted and meet the demand to do too much in too little time. The media not only create a culture of attention deficit disorder but one of hyperactivity as well.

Attention deficit disorder and hyperactivity are only the extreme manifestations of our internalization of the rapid movement of media images and sounds. Jane Healy has studied both disorders and sees their relationship to heavy television viewing and prolonged playing of video and computer games.[63] Attention deficit disorder is independent of hyperactivity, but they sometimes occur together. Both have been diagnosed as forms of learning disability. Attention deficit disorder entails the inability to maintain attention and concentrate on a specific task. Hyperactivity applies to those who cannot concentrate and also have trouble sitting still; their bodies must be in constant motion. Healy cites research on the brain and central nervous system suggesting that overexposure to the mass media and computer may prevent synapses from forming properly in the left hemisphere of the brain so that the ability to read, write, and think is stunted.[64] Someone who lacks facility in language does not understand much of what goes on in school and experiences wandering attention. This mimics the visual media, which moves rapidly from one image to another, from one topic to another.

The ability to read, write, and think is generally declining. The problem is not merely too much television or too much video and computer games but the low quality of conversation and of what is read. As Healy notes, students are not expected often enough to read complex sentences and struggle with

60. Carr, *Shallows*.
61. Carr, *Shallows*, 142.
62. Jackson, *Distracted*; Carr, *Shallows*.
63. Healy, *Endangered Minds*.
64. Healy, *Endangered Minds*.

complex thoughts.[65] Moreover, there has been a dramatic decline in the level of discourse in the media, in textbooks, and in the books most adults read. Given the fast pace of life in a technological society and the rapid movement of images in the media and video and computer games, it is understandable that we have all become mildly hyperactive and all suffer from attention deficit disorder to some extent. Hyperactivity and attention deficit disorder are this century's stimulus shields.

The impact of the media (including the computer) includes loss of memory, personal thought, and feeling. Memory is an integral part of one's existence; it is a product of one's relationship to reality defined by culture and placed into the context of one's entire life. Memory tends to be centered on key events and people that have acquired a certain symbolic quality. There is neither past nor future in the media, only the present. The media aestheticizes reality and turns life into a spectacle to be enjoyed now. There is no need for memory, at least long-term memory, when one lives exclusively in and for the moment.

Visual images have a largely emotional impact on us. When we watch television, the activity in the left hemisphere of the brain is about what it is when we are asleep. Today adults as well as children spend much more time watching the media and playing video and computer games than they do reading. What reading they do tends not to be serious. The result is catastrophic. Healy observes that "professionals report that children in classrooms seem to be thinking and learning in increasingly more nonsequential and visual ways."[66] Consequently, there is a decline in logical thought and critical reflection. The image-oriented individual is more readily influenced by propaganda, advertising, and public relations.

Healy reports that the use of music with a heavy beat "blocks the capacity for thinking."[67] Music is so pervasive that it is part of our environment. Silence is prohibited. The music-addicted mind functions as a stimulus shield.

Finally, Brod discusses the anxiety that the computer creates for its users. They tend to make one of two responses: a "struggle to accept the computer" or "overidentification" with it. Those who embrace the computer and make it the center of their lives suffer the "loss of the capacity to feel and to relate to others."[68] In other words, they become like the computer.

65. Healy, *Endangered Minds*.
66. Healy, *Endangered Minds*, 58.
67. Healy, *Endangered Minds*, 175.
68. Brod, *Technostress*, 16–18.

The mass media and the computer have led to a loss of memory, thought, and feeling. We are becoming more like machines, which possess none of these capabilities. Our stimulus shield allows us to internalize technological stimuli so that we act according to reflex, not personal thought.[69]

The form of the media requires a stimulus shield so that we can more readily adapt to the constant stimulation of our senses. The content of the media, sociological propaganda, plays a part as well in the recreation of the human being. Technological utopianism creates an aesthetical reality dominated by the pursuit of happiness. For the individual happiness defined in terms of total consumption is the paramount purpose of technology. The media both in form, visual images, and in content, technological utopianism, aestheticizes life. It is no wonder then human relationships are largely aesthetical. This, of course, is how children approach life. Public opinion is based on desire, and it unifies society as envy (the negative unity of the negative reciprocity of envy). Children openly show their envy to the extent that they have not learned self-control. The content of the media preaches conformity in that the peer group and its value of fitting in are central themes. Visual images provide ersatz norms of style, which are nothing more than public opinion. The peer group contextualizes public opinion for the lonely and fearful individual.

The creation of the media: a distracted, hyperactive, pleasure-seeking, envious, emotion-dominated child, who acts without thinking, is only capable of shallow relationships, and is lonely with a need to conform to the peer group. This creation is the opposite of the ideal human in Western civilization: an *individual* who employs *reason* in the service of *freedom*. We have become the childish machines we created: the media creates us in its image. In this way the technological system establishes total control over us: it defines what it means to be human.

---

69. Ellul, *Technological Society.*

Chapter 3

The Technological Personality

Originally published in the *Bulletin of Science, Technology and Society* 24, no. 6 (2004) 488–99.

The objectification of experience in technology, especially the mass media, has had a profound influence on the modern personality. Personality is now characterized by two dialectics: the objective and the subjective and the inner and the outer. This is in sharp contrast to pre-eighteenth-century personality for which the distinction between objective and subjective and inner and outer was much less pronounced.

The objectification of the personality is aided by the loss of judgment and responsibility. Human technique obviates the need for judgment, decisions, and responsibility. Consequently, the individual is extremely vulnerable to those objectivizing tendencies. Technology colonizes our experiences, opinions, emotions, and consciousness.

Just as technology leaves us with secondhand experiences, it provides us with secondhand opinions. Public opinion is based on secondhand knowledge rather than experience or readily intelligible facts. Public opinion is a necessary component of a mass society; it is a substitute for interpersonal knowledge that arises from practical, everyday life in the context of intense familial and community relations. The mass media provide us with ersatz experiences on which opinion rests. If personal knowledge is phenomenal, public opinion is epiphenomenal; it is concerned with what most of us know little about through experience or serious thought.[1] Therefore, public opinion is secondary both in regard to its source—the mass media—and its type of knowledge—epiphenomenal. More obvious than the artificial character of public opinion is its fragmentation.[2] There is no coherence or continuity to the totality of opinions. A

1. Boorstin, *Democracy and Its Discontents*.
2. Boorstin, *Democracy and Its Discontents*.

sure sign of fragmentation is the inconsistent (if not contradictory) quality of many opinions. For example, American opinion supports preserving the environment and economic growth simultaneously. Opinions are always out of context, for issues are presented to the public consecutively as self-contained entities; they resemble facts that appear to be autonomous. The fragmentation of opinion reflects the fragmentation of information conveyed through the media.[3]

Partly as a consequence of fragmentation, public opinion is based on simplified issues.[4] Complex and profound problems can only be resolved, and even then rarely, through long and arduous discussion. Because opinion forms (or is pressured to form) on an enormous number of issues, many of which are outside of most people's personal expertise, the media must simplify the issue. Only then can public opinion emerge. Criticism of American politicians, for example, sometimes centers on their tendency to simplify issues when the problem is actually much more a result of the mass media, which can only operate profitably through simplification, and the public, which, no matter how well educated, due to technical specialization, is ignorant about most issues.

If public opinion is knowledge (albeit simplified and secondary) at one level, it is desire (and fear) at another. Public opinion is ultimately about that which one desires to occur or wishes to be eliminated or controlled. This is evident in the very terms under which opinion is *solicited—agree* or *disagree, yes* or *no, approve* or *disapprove, like* or *dislike.* Public opinion tends to follow the desire that technology's promise unleashes and advertising presents to us. Therefore, public opinion is decidedly ephemeral.[5] Its norms change as everyday life changes and new objects of desire emerge.

Secondhand opinion is a hybrid—part superficial knowledge and part desire. Secondhand experience and secondhand opinion result in a loss of a sense of reality. Heavy television viewers experience the greatest distortion of reality.[6] The loss of a common symbolic reality, which only language can provide, necessarily results in the loss of a sense of reality. Those who are dominated by visual images have a false sense that reality is on television and in the computer—the two sources of most of our information.[7]

Secondhand emotions originate in the media. The media, which provide us with vicarious experiences, are dominated by visual images whose

3. Miller, *Boxed In*; Postman, "The News."

4. Ellul, *Propaganda.*

5. Ellul, *Political Illusion.*

6. Gerbner and Gross, "Living with Television."

7. Stivers, *Technology as Magic.*

fundamental appeal is emotional. Consequently, in viewing television programs, movies, and computer videos, we experience emotions that are shallow and transient. Moreover, they become what Arnold Gehlen has called
"emotional schemata," or emotional stereotypes.[8] Emotionally, we are like
Pavlov's dogs responding to emotional conditioning.

When a subjectivity based on lived experience is emptied out and
replaced with objective images, the personal and impersonal are inverted.
In recent years, people have grown more willing to reveal intimate details of their personal lives to casual acquaintances or even strangers. The
model for this, of course, is the talk show where ordinary people along
with celebrities proclaim that they once were drug addicts, experienced
child abuse, or suffered from illiteracy. When we share intimate details
of our lives with acquaintances and strangers, our discourse resembles
the *anonymous* discourse of the mass media. It is no longer important to
whom we address our remarks.

Even as we divest ourselves of our personal lives, we attempt to become
personal by identifying with celebrities and even products on television. There
are a couple of reasons for this. One is that reality is in the media, and the
celebrities have lives infinitely more interesting than our own humdrum lives.
Another is that we know more people in the media than in real life, and we
know them more intimately.[9] Most of our relationships except for family and
close friendships are superficial. Indeed, research indicates that we feel closer
to television personalities than we do to acquaintances, and the strength of
this feeling is a close second to what we feel for our friends.[10]

We come alive and become interesting by living through and resem-
bling these celebrities and other advertised products. We are thus objectified as images. What prevents everyone from becoming the same image
object is the myriad of ways of assembling one's total image. This totality is
always changing, because the images we live in and through can readily be
substituted for others. Life as represented on television and the other media
is an experimental life.

Consciousness usually contains personal experience and emotion,
but technical consciousness does not. It reduces every activity and relationship to an instrumental one and turns knowledge into abstract information. Technical rationality makes up an ever-increasing part of individual
consciousness. We become full-time consumers of technical information
both at work and at leisure. The excessive information of the computer and

8. Gehlen, *Man in the Age*, 78–79.
9. Martin, *Who Am I?*
10. Leroux, "Our Electronic Friends."

the media, Ellul noted, turns us into "exclusive consumers," with the result that even shopping for and acquiring consumer goods becomes a means of gathering and using information.[11] That is, we become connoisseurs of technical products. Even if one only owns an inexpensive and simple music system, one can become a connoisseur by gathering sophisticated information about speakers.

Technical rationality is collective and impersonal; it does not depend on experience, practical knowledge, and moral judgment. Moreover, the growth of technical rationality implies a fractionalized consciousness. This involves not only the split between collective rationality and subjective reason but also the fragmentation within technical rationality itself, because of specialization.[12] Technical information is unintegrated culturally and therefore proliferates wildly and appears to be random. Information is never neutral; it directs our behavior, including moral conduct. The most important information in traditional societies was practical knowledge useful for group survival that was given symbolic meaning. Information about hunting, for example, described traditional ways of hunting and explained how the deities or their representative, the master of the hunt, provided the animals, which humans had to respect even as they hunted. Practical knowledge was thus ritualized because it was set in a narrative about the meaning of hunting.

Today, information tends to go in opposite but mutually reinforcing directions. Knowledge has become abstract, both as theoretical knowledge and technical knowledge.[13] Technical knowledge tends to supplant tradition or common experience in providing information that is practical in the sense of being efficient, objective (decontextualized), and meaningless. Such abstract information is not sufficient to meet human emotional needs. Traditional information did this by providing a symbolic context for practical knowledge. As shared meaning declines in a technological civilization, knowledge must develop that is emotional as an end in itself rather than as an epiphenomenon of shared meaning. Emotional knowledge becomes a compensation for abstract knowledge, especially technical knowledge. The most prevalent sources of emotional information today are the televised and the computerized visual image.

How does the individual respond to the reification of experience, knowledge (opinion), emotion, and consciousness? In traditional societies, much of what is today considered objective knowledge (technical

---

11. Ellul, *Technological Bluff*, 330.

12. Kahler, *Tower and the Abyss*.

13. Ellul, *Technological Bluff*.

knowledge) was a kind of symbolic, intersubjective knowledge. This practical knowledge was embedded in a cultural matrix that narrowed the gap between subjective and intersubjective experience. Objectivity becomes a problem when it is cut loose from symbolic meaning, for then it appears to be inorganic. The individual intuitively feels menaced, indeed overwhelmed, by a nonliving objectivity. As Owen Barfield has observed, throughout most of history, humans identified objectivity with organic nature—a living whole that was experienced as having an inward consciousness.[14] By symbolizing nature, humans participated in creating and maintaining it. Objectivity was experienced as intersubjectivity. Because technology at a certain threshold weakens, if not destroys, symbolic meaning, today the objective and subjective are at odds. That is, human subjectivity is a reaction against the objectification of human experience, knowledge, and emotion, but a reaction that does not permit escape.

Technical consciousness is an abstract, impersonal consciousness. This cold consciousness, Kahler argued, has contradictory consequences: greater insensibility and greater sensibility.[15] The former refers to depersonalization or detachment. Insensibility to another's plight or even my own means that I analytically search for the causes of the misfortune and what remedies there are for it instead of responding intuitively, emotionally, and with moral judgment. In so doing, we deaden ourselves against suffering and become indifferent to others even though we may urge political action or make a contribution to charity.

At the same time, however, technical consciousness produces a greater sensibility to visual stimuli. Alfred Crosby has demonstrated that prior to the late Middle Ages, Western civilization had relied more on the senses of touch and hearing than on sight. But beginning in the thirteenth century, an emphasis on the synthesis of visualization and quantification led to the technological mastery of life.[16] Today, we live in a world in which visual images are dominant and proliferating rapidly. We have become connoisseurs of visual details, not those necessary for survival as with a hunter-gatherer group but those consumed for the aesthetic pleasure of the detail. We also become more sensitive to the personalities of others and acquire a hyperreflexivity about ourselves. Musil, in *The Man Without Qualities*, perfectly illustrated the new insensibility and sensibility.[17]

14. Barfield, *Saving the Appearances*.
15. Kahler, *Tower and the Abyss*.
16. Crosby, *Measure of Reality*.
17. Musil, *Man Without Qualities*.

The new sensibility reinforces the tendency to differentiate one's personality from that of others. Georg Simmel observed in 1900 the tendency of people to emphasize accidental details such as hairstyle as important aspects of personality.[18] The reason he gave is the standardization affected by the "social-technological mechanism." The money economy, bureaucracy, and the political state create rules that make citizens increasingly similar. To struggle against this standardization, the individual emphasizes the aesthetic details of the self that have come to constitute personality.

The loss of direct experience results in a loss of a sense of reality. In a technological civilization, a shared symbolic reality gives way to a reified reality—in the media and the computer. Reality is always somewhere else. The individual compensates by creating her own imaginary world. An apt example is a factory worker who performs boring, repetitive tasks while living in a world of daydream and fantasy. A meaningless existence demands an imaginary world. The loss of a feeling of reality likewise unleashes desire, which is the mechanism of fantasy. Advertising preys upon this vulnerability by creating a fantasy world of desirable products. Moreover, this world of desire and fantasy overflows the boundaries of the self into human relationships. We end up expressing absurd and extravagant opinions on matters we know little about.[19] With technology so dominant, subjectivity is narrowed to human instinct.[20] At least my instincts, such as aggression and sexuality, belong to me.

Loneliness at the Core of the Technological Personality

At the inner core of this new subjectivity are loneliness and a fear of others. Behind these feelings are a sense of meaninglessness and a sense of powerlessness. Technology appeals to our will to power. It provides us with a false sense of power: technology's (the automobile, the computer) power is my power. In contradictory fashion, however, it contributes to a confused sense of impotence, because that power is really not my power. Humans cannot keep up with the speed, power, and information processing ability of machines, computers, and technological systems. Moreover, the increase in information about technological means and technological objects may lead to a failure to make a choice. Too many choices and too much information about them can create a situation of sensory and

18. Simmel, "Metropolis and Mental Life."
19. Gehlen, *Man in the Age.*
20. Stivers, *Culture of Cynicism.*

intellectual overload with a resultant feeling of impotence. The pervasive but tacit sense of powerlessness today is a reaction to the overwhelming and universal power of technology.

Feelings of powerlessness and meaninglessness are mutually reinforcing. There is an inverse relationship not only between power and moral values (a type of symbolic meaning) but also between technical information and symbolic meaning. The excess of information in a technological age results in a broken vision of the world. One loses a sense of the past, the ability to evaluate the present in terms of the past. The triumph of the computer and the visual images of the media produce a "culture of forgetting." Studies indicate that our memories decline to the extent that everything is done for us (the computer) and shown to us (the images of the media).[21] Memory centers on events that have acquired emotional, intellectual, or moral significance to us. The past understood in either mythological or historical terms acquired meaning through key events that took on symbolic meaning. The information from the computer, as already noted, is abstract, logical, and quantitative but ultimately meaningless. The cynical worldview of the computer is that we live in a random and meaningless world about which the omnipotent computer can generate an infinite amount of information that we can exploit to our advantage.

Symbolic meaning, by contrast, provided humans with a measure of power. Historically, symbolization was a way for humans to gain control over their milieu. By enveloping it within their symbolic system, even as they were acknowledging its greater power, they gained a sense of mastery. Ellul argued that the human ability to symbolize has been the single most important factor in the cultural evolution of the human race.[22] Because technology is our own creation, we do not perceive a need to symbolize it in the traditional sense. It is only when we confront a foreign power—nature— that we bring it within our symbolic net. Symbolization simultaneously creates a meaningful world and allows us to *distance* ourselves from this world. Without effective symbolization, we have no way to keep technology from invading and conquering culture.

The rise of technology and the decline of a common morality involve the erosion of moral meaning and a tacit feeling of powerlessness. Feelings of meaninglessness and powerlessness extend beyond technology to human relationships. With the loss of a shared morality and the proliferation of techniques to control others, we are freer to manipulate and exploit others. Of course, they are just as free to do the same to us. The destruction of

21. Ellul, *Technological Bluff*, 330.
22. Ellul, "Symbolic Function."

a common morality means that we are not morally obligated to help others, nor can we expect assistance from them. Without a minimal level of trust that a common morality provides, human relationships become vague and dangerous.[23] When human relationships become uncertain, competitive, and potentially hostile, the individual protects herself by failing to express her true feelings to someone who, in all likelihood, is concealing his motives as well. Human relationships become an arena in which individuals play a number of roles, feign certain emotions, and manipulate others to their advantage. To be alone, however, is not necessarily to be lonely. Loneliness arises in a context of meaninglessness and powerlessness, and it brings with it a fear of others.

There is a way to describe modern subjectivity: puerilism. If modern technology tends to diminish normative reason, responsibility, personal experience, and moral judgment, then it renders us childlike. J. H. Huizinga has provided important insights about modern puerilism.[24] The paramount characteristic of puerilism is the inversion of work (and every serious activity) and play. We treat play (e.g., sports, television programs, video games, and entertainment in general) with a seriousness that borders on fanaticism. Witness the growing fervor of sports spectators all over the world and the amount of television coverage given to sports. Concomitantly, we regard the serious activities of life, such as business and politics, as a game whose sole purpose is to gain personal advantage over others. The global economy has institutionalized business as a game.

Huizinga observed that the sentimental idealization of youth, so prevalent in the mass media, is the work of adults.[25] The so-called youth culture is the expression of a technological utopianism that extols the health and happiness of the human body.[26] Both in terms of physical development and freedom from responsibility, youth symbolizes these cultural values. Examples of puerilism include "the need for banal entertainment" and the "search for gross sensations," which are characteristic of a state of "permanent adolescence."[27] This is akin to Robert Pattison's definition of the vulgar in the context of rock music: "common, noisy, crass, and untranscendent."[28] The vulgarian simply follows his instincts.

23. van den Berg, *Changing Nature of Man*.

24. Huizinga, *Shadow of Tomorrow*.

25. Huizinga, *Shadow of Tomorrow*.

26. Stivers, *Culture of Cynicism*.

27. Huizinga, *Shadow of Tomorrow*, 170–82.

28. Pattison, *Triumph of Vulgarity*, 6–7.

Communication, too, has become vulgar. The widespread use of slogans, intolerance toward the opinion of others, excessive flattery, and the expression of unbridled instincts are all signs of puerilism.[29] Mass entertainment is for the most part easily recognizable as puerile. Comedy has become gross, cruel, and stupid. Beavis and Butthead and Tom Green are examples that readily come to mind. Sports participants and their fans seem to compete to see which group can be more juvenile. Trash talking, finger pointing, the wave, chanting, loud booing, the intimidation of referees, and fights on the field and in the stands are not the actions of adults.

The modern personality is puerile in its subjectivity, but the inner core of this subjectivity implies that the child is lonely and fearful. Just as the decline of a common, symbolic intersubjectivity created a tension between the objective and the subjective (the latter a compensation for the former), so, too, did it effect a contradiction between the inner and outer dimensions of personality. The outer surface of the technological personality, seemingly, is the opposite of the lonely child but is a compensation and cover for it.

The outer surface of the personality is flexible, adjusted, outgoing (friendly and garrulous), confident, and mildly hyperactive. This is the personality modern organizations require and the one that is represented in the media in the form of the celebrity.[30] Flexibility means that one can and will play any role one is given. Good role players engage in deep acting. Surface acting, according to Arlie Hochschild, entails feigning certain feelings toward someone with full consciousness that one feels otherwise; deep acting involves a more complete assimilation of the role to the self (at least for the duration of the scene) through repeated commands to the self (a kind of self-hypnosis) or by imaginatively constructing the desired emotions internally.[31] Deep acting is more tiring because it temporarily requires a greater transformation of the self, but it is more convincing.

Adjustment suggests that one is a team player, one who fits in. The adjusted person makes a home in the group or organization and more than anything else fears rejection. Together, flexibility and adjustment create the obedience-to-authority mind-set that enables one to simply follow.

In his classic *The Organization Man*, William Whyte suggested that managers were expected to be outgoing even in private life in the suburb.[32] The outgoing, friendly, talkative, and confident individual has a better chance of manipulating customers, employees, and neighbors than the introverted

29. Huizinga, *Shadow of Tomorrow*.
30. Jackall, *Moral Mazes*.
31. Hochschild, *Managed Heart*.
32. Whyte, *Organization Man*.

person. Because we are all expected to be like this, we demand this of each other; we do not like diffident and reserved people.

With the decline of a common morality, every relationship becomes one of power: *adjust* to greater power and *manipulate* lesser power.[33] The technological personality contains both dimensions. It stretches from one pole to the opposite so that any individual's personality falls somewhere in between in terms of emphasis. Some of us are more adjustment oriented, whereas others are more manipulation oriented. Flexibility allows one to fit in, whereas being outgoing, friendly, confident, and talkative facilitates the manipulation of others. The model for the technological personality is in the media—the celebrity.

No one has understood the celebrity's relationship to television better than George Trow.[34] The celebrity is the personification of television. The *cold child* is Trow's name for television. On television, his home, the celebrity has a short attention span and is self-indulgent, like a child. Moreover, he exudes pseudo-cheerfulness and pseudo-intimacy, but it is a cold or false cheerfulness and intimacy. The pseudo-cheerfulness and pseudo-intimacy are anonymously directed to anyone watching—the potential consumer. They are cold, manipulative emotions. The celebrity links television with its audience by providing us with vicarious experiences and feelings of love. This love, however, turns out to be nothing more than familiarity. The celebrity is, of course, associated with products so that the latter come to embody the qualities of the former. As Trow noted, "The most successful celebrities are products. Consider the role in American life of Coca-Cola. Is any man as well loved as the soft drink is?"[35] No wonder we wear our clothes as advertisements.

What is the relationship between the inner core of the modern personality—the lonely and fearful child—and the outer surface—the outgoing and confident child? The head of the outgoing, confident child rests on the body of the lonely, fearful child. The former is a comment on and compensation for the latter. Van den Berg has observed that modern people talk for the sake of talking. It is of no interest to them what the talk is about.[36] Who better fits this description than the celebrity? Trow suggested that television is about pseudo-intimacy and that only celebrities have dual lives—an intimate life and a lonely everyday life in a mass society.[37] Only celebrities

33. Stivers, *Culture of Cynicism.*

34. Trow, *No Context.*

35. Trow, *No Context,* 48.

36. van den Berg, "What is Psychotherapy?"

37. Trow, *No Context.*

appear to be fully human. Getting to know them appears to be an end to a mild but endemic loneliness.

This consideration of the two dialectics of the technological personality—from the objective to the subjective and from the inner to the outer—suggests that there is little genuine individuality left to the modern self. The technological personality is a creation of modern organizations and the mass media. The technological personality is the self that a technological civilization demands and creates.

The Tempo of Society as Stress

Modern technology not only helps create a personality type but also generates a great amount of stress. Stress is a universal phenomenon, but its forms and the human response to it are culturally and historically specific. The main issue here is how the technological personality confronts various kinds of stress and eventually internalizes them.

Stress is a general term that refers to almost any negative aspect of life. I will use it to refer to aspects of our physical and social environments that threaten our comfort or well-being and *demand* a response.

There is a growing body of research and theory on environmental stress.[38] It is not my purpose to clarify and resolve the different interpretations of stress. Some approaches emphasize its physiological aspect; others focus on its psychological dimension.

Some think of stress in the context of brain activity, others in the context of the ability to process information, and still others in the context of systems theory (the relation of the demands of the environment to the individual's ability to control the environment).[39] Despite the variety of perspectives on stress, there are some common insights and categories in its study.

The earliest sustained studies of stress were about physiological responses to physiological challenges in the context of disease. The pathogen model of stress helped call attention to the consequences of long-term exposure to stress: exhaustion and a lower resistance to disease. In this view, the problem is not mere stress but its persistence.[40]

Increasing attention has been devoted to psychological stress as a factor that accompanies physiological stress or as a factor in itself.[41] In the case of potential environmental stress, the individual's interpretation of the situation

38. Evans, ed., *Environmental Stress*; Vanderburg, *Labyrinth of Technology*.
39. Evans, ed., *Environmental Stress*.
40. Baum, Singer, and Baum, "Stress and the Environment," 16–17.
41. Baum, Singer, and Baum, "Stress and the Environment," 17.

is its psychological mediation. Is the event (source of the stress) perceived to be threatening, harmful, or just challenging? Moreover, what is the meaning of the potentially stressful event? For example, one man's music is another man's noise. Experience plays a major role in psychological stress. Personal knowledge about the detrimental effects of long-term exposure to loud noise may increase the level of psychological stress one experiences in such a situation, for instance. Furthermore, some sources of stress in the environment possess symbolic meaning. Loud music of certain types may come to symbolize unruly teenagers who are themselves perceived to be a source of stress. The perception of whether one can control the source of stress also has some impact on how stress is experienced.[42]

The experience of stress leads to coping behavior. How does one respond to the stress? There appear to be two principal ways. One is manipulation—for example, the individual either removes the stress or leaves the scene of the stress. The other is adjustment—for example, the individual attempts to change her internal environment. Drugs, alcohol, meditation, and relaxation exercises may distract one from the stress.[43]

Psychological stress may continue well after the source of the stress has been attenuated. Moreover, chronic or repeated stress may produce an effect that is a consequence of continuously responding to stress—aftereffect. There are both physiological and psychological aftereffects. Psychological aftereffects may include "decreases in cognitive functioning and reduced tolerance for frustration, aggressiveness, helplessness, decreased sensitivity to others, and withdrawal."[44] Perhaps the most pervasive type of stress today is the tempo of society.

Tempo is a musical term that designates the rate of speed at which a composition is played.[45] It sometimes refers to the speed at which life is led. The term "pace of life" is often substituted for *tempo of life*, but Levine maintained that the former term encompasses more than the latter. Pace of life includes speed as well as the "tangled arrangement of cadences" and "perpetually changing rhythms and sequences, stresses and calms, cycles and spikes."[46] Pace of life encompasses social time in its entirety. Because most people experience social time in terms of tempo I will use the term *tempo* interchangeably with *pace of life*.[47]

42. Baum, Singer, and Baum, "Stress and the Environment."
43. Baum, Singer, and Baum, "Stress and the Environment," 20–21.
44. Baum, Singer, and Baum, "Stress and the Environment," 34.
45. Levine, *Geography of Time*, 3.
46. Levine, *Geography of Time*, 25.
47. Levine, *Geography of Time*, 24.

Like other animals, humans have bodily functions and processes that are timed to nature, the circadian day, and seasonal cycles.[48] Industrialization altered the natural rhythm of human existence, first at work and later at leisure, as people were forced to keep pace with the machine. Today, the computer imposes its time—the nanosecond—on people. Obviously, the tempo of life is accelerating.

Research on the personal experience of tempo has identified several categories—first, time urgency. This may include a preoccupation with time, rushed speech, rushed eating, driving too fast and impatiently, constantly making lists and schedules, impatience when waiting, and feeling irritable or bored with nothing to do. Time urgency entails a compulsion to do as many things as rapidly as possible. We seem to suffer from a sense of time scarcity.[49] Other aspects of tempo include the perceived speed of the workplace and the speed outside the workplace—for example, family, leisure, and the preferred level of activity in life as a whole.[50] Modern technological societies can be characterized as high tempo. Technology has shrunk time and space, thereby permitting us to live and work ever more quickly. Not all of this speed is experienced as stress, however, as we will see below.

The tempo of television commercials, television programs, and films has accelerated.[51] Commercials contain the most hyperactivated imagery: There are an average of ten to fifteen image changes per thirty-second commercial and seven to ten per sixty-second segment during a typical television program.[52] The tempo of films has likewise increased. Many today complain about slow-moving films that feature dialogue and character development. The faster and more frenzied the action, the better. Events and time become accelerated, condensed, and thus simplified. Consequently, the media are the great equalizers. Between television and the computer, amusement is conducted at an ever-accelerating pace. Because the media produce ersatz norms, especially for heavy users, there is a sense that to avoid boredom, life outside the media should be fast paced.[53]

Consumption, which is enforced and must keep up with production, is also accelerating. Economist Steffan Linder identified three forms that the acceleration of consumption assumes. The first is simultaneous consumption, which is when an individual uses more than one consumer

48. Rifkin, *End of Work*.
49. Linder, *Harried Leisure Class*.
50. Levine, *Geography of Time*, 97.
51. Gleick, *Faster*.
52. Mander, *Absence of the Sacred*, 85.
53. Ellul, *Humiliation of the Word*.

product simultaneously—for example, reading, watching television (even several programs), listening to music, and carrying on a conversation. The second form is successive consumption, which refers to using products or services consecutively but decreasing the time spent on each activity. For example, instead of playing eighteen holes of golf, one plays nine holes and then immediately goes fishing for several hours. The third form involves substituting a more expensive version of a consumer product for a less expensive version in response to advertising intended to turn us into connoisseurs, first at the level of information and then at the level of experience.[54] The three forms are interrelated and derive from the cultural belief that the meaning of life resides in consumption.

Perhaps there is no better example of the acceleration of consumption than tourism.[55] Travel in the nineteenth century was leisurely and often undertaken for economic, political, and cultural reasons. By the second half of the twentieth century, tourism, which has little in common with travel, had become a major industry. Tourism mastered successive consumption by packaging as many tourist sites and countries into the shortest amount of time possible; moreover, tourist enterprises provided consumers with simultaneous activities on the tour bus or ship—for instance, food, music, lectures, movies, and so forth. The goal is to persuade the consumer to upgrade his tour next time with an even better product.

The acceleration of family life has received less attention than that of work until fairly recently. Hochschild's insightful study of how families manage time urgency is especially helpful, because she looked at work and family concurrently.[56] She argued that work has become more humanized, whereas the home has become more industrialized. More and more parents try to become efficiency experts at home to meet the demands of marriage and family life. Of course, both parents working exacerbates the situation. The number of organized activities outside the home that children participate in has been increasing. Hochschild termed this *domestic outsourcing*: recreation, counseling, tutoring, music lessons, sports, and meals often occur away from the home. Consequently, dutiful parents act as travel agents and cab drivers for their children.

The consciousness of time scarcity in the family is expressed in the term *quality time*. Although parents spend only limited time with their children, it can be so intense and meaningful that the quality of their time together more than compensates for its brevity. One problem is that quality time lends itself

54. Linder, *Harried Leisure Class*, 79.
55. Boorstin, *Image*.
56. Hochschild, *Time Bind*.

to superficial amusement instead of serious discussion, which takes more time than quality time permits. Hence, quality time becomes a reward that guilty parents bestow on their children.[57]

The hectic pace of life that children experience reaches its culmination in college. David Brooks wrote about students at Princeton University who are so busy that they have to make appointments to see their friends.[58] The proliferation of external, organized activities in childhood is apt preparation for college.

Related to the tempo of life is the experience of simultaneity, which entails the compression of quantitative time and the loss of qualitative time. It is a long-standing cliché that our lives are ruled by the clock. Following Benjamin Franklin, we know that time is money and that punctuality is a virtue. Moreover, we suffer from time scarcity in modern societies because we spend so much time at work, shopping, servicing our possessions, and at leisure.[59] The latter (leisure) maintains the hectic pace of the former (work).

The computer creates a synchronous society by reducing time to ever-smaller units. The computer produces the experience of the instantaneous as it processes increasingly enormous amounts of information in nanoseconds. As Ellul observed, "Real time, in which the computer now functions, is a time looped in advance and made instantaneous."[60] With the assistance of the computer, technology programs time—that is, it subordinates time to its own functioning rather than to human existence. As such, it reduces the past and future to the present. But because the present is shrinking to the nanosecond, technology appears to exist outside of time.

In addition to reducing quantitative time almost to the zero point, technology destroys qualitative time. Technology makes tradition, the memory of the past, and experience irrelevant, for it objectifies experience. The past ceases to be a living past and exists merely as history. A living past is organized around symbolic events that serve as key elements of a group's identity.

Likewise, technology makes the future as the meaning of history superfluous. A qualitative future requires that history be open, not predetermined in the process by which the future is realized. But technology attempts to control the future by eliminating it—that is, by creating the eternal present of

57. Hochschild, *Time Bind.*

58. Brooks, "Organizational Kid."

59. Linder, *Harried Leisure Class.*

60. Ellul, *Technological Bluff,* 95.

a technological utopia.[61] The myth of progress promises a future that, despite its accidental differences, is only a repetition of the present.[62]

What are the consequences of a faster tempo of life that borders on an experience of simultaneity? First, there is a decline in memory. The more we are forced to live almost exclusively in the present, the more the past fades from memory. We only remember what is important to us, and increasingly what is important is reduced to my private existence. Second, the quality of decisions suffers.[63] Reflection cannot be hurried; it requires time to integrate information, see relationships, and make judgments. As Ellul has noted repeatedly, a technological civilization demands reflex, not reflection. Decisions tend to be automatic in furthering the cause of technological growth. Critical reason, which requires time and solitude, becomes scarce in a technological civilization. Third, there appears to be a decline of pleasure.[64] To be distinctly pleasurable, an event needs to be experienced in a relaxed, not a hurried, fashion. Our harried existence reduces the opportunity for intense pleasure. Quantity replaces quality.

If pleasure (or at least conscious pleasure) is declining, ecstasy is on the upswing. Ecstasy is an altered state of consciousness. Mysticism is a form of ecstatic experience, as is the high that results from frenzied activity (e.g., long, intense exercise; loud, repetitive music). Ellul argued that much ecstasy today is "a function of the acceleration of the tempo of the technical society."[65] Echoing this, Milan Kundera mentioned that "speed is the form of ecstasy the technical revolution has bestowed on man."[66] Perhaps he should have said that speed is the main cause of ecstasy today, for to act by reflex for long periods results in a state of ecstasy.

Tempo alone is not responsible for the widespread experience of ecstasy; it is also a consequence of the regimentation of existence. As bureaucratic and technical rules proliferate, humans sense that they have lost personal control over their lives. Escaping into the realm of ecstasy seems to be rebellion against a too rational, too orderly existence.[67] Drugs, alcohol, sex, violence, music, and video games are all forms of ecstatic escape. Technology therefore directly produces ecstasy through the tempo it imposes on society and indirectly produces it as compensation for regimentation. Some of the

61. Ellul, *Technological System*.
62. Horkheimer and Adorno, *Dialectic of Enlightenment*.
63. Linder, *Harried Leisure Class*.
64. Kerr, *Decline of Pleasure*.
65. Ellul, *Technological Society*, 421.
66. Kundera, *Slowness*, 2.
67. Ellul, *Technological Society*.

compensations for the impact of technology, for example, computer games, mimic and thus reinforce the tempo of a technological civilization.

The Technological Personality as Stimulus Shield

The consequences of stress are mitigated to a certain extent by the technological personality. Humans have learned to internalize some stressful stimuli and, in so doing, have adapted to them. The concept of the *stimulus shield* as applied to technology is so important to my argument that we will look at Wolfgang Schivelbusch's seminal idea in some detail.[68]

The theme of his book *The Railway Journey* is how the railroad, specifically travel by rail, affected the perception of time and space. In his view, this is only one part of how industrialization in the nineteenth century influenced the human psyche. Schivelbusch demonstrated how railway travel created panoramic vision. The velocity of the train made it impossible for the traveler to intensely study the landscape. Now perception became extensive: one viewed as much as possible in the brief time allotted by the speed of the train. One learned to take in the entire landscape at once rather than consecutively.

Schivelbusch's discussion of the stimulus shield as a way of normalizing a technology that produces fear and anxiety allows us to see stress in a new light. Railway travelers, he argued, created new ways of perceiving and behaving that allowed them to forget their fear of train travel. The size and speed of trains frightened people; accidents appeared more imminent and more deadly. The train manufacturers tried to mitigate the fear that trains would crash by minimizing the jolts and noise that suggested the possibility of a train wreck; moreover, the compartment decor created an elegant atmosphere that distracted travelers from their anxiety about a train wreck. Reading on trains almost became mandatory; it distracted passengers from their fear. Finally, panoramic vision represented a way of internalizing the speed of the train, which suggested a possible train wreck.

Schivelbusch explained that the higher the technology, the more catastrophic its destruction. Historically, catastrophes were the result of natural forces, such as a flood, which could destroy human products from the *outside*. A higher level technological object, like a train, is destroyed from the *inside* by means of its own power. Because natural catastrophes were rare occurrences and external to humans, nature was not as frightening. Higher level technologies, however, were catastrophes waiting to happen. It is one thing to be riding in a horse and buggy with the distant possibility of a tornado

68. Schivelbusch, *Railway Journey*.

and quite another to be riding on a train that could at any time explode. Schivelbusch also noted that the higher the level of the technology, the more denaturalized human consciousness becomes. When that denaturalized consciousness collapses in the case of a technological accident, the psychic shock is much greater than in the instance of a natural catastrophe.

The denaturalized consciousness includes the idea of the stimulus shield, a term taken from Freud. It is more or less synonymous, Schivelbusch argued, with Simmel's term *intelligence*.[69] For Freud, new and strong stimuli, such as train travel or flying in an airplane, tend to produce a "skin layer of consciousness" or a "thicker layer of skin" so that the stimuli become less threatening. In other words, the observer internalizes the stimuli, becomes acclimated to them, and dulls them so that it would take novel stimuli to once again frighten him. Simmel's comparable idea is that when faced with pervasive and intense stimuli, such as living and working in a large city, the individual reacts intellectually rather than emotionally to them. This emotional indifference is a way of protecting the individual from emotional upheaval and exhaustion. The panoramic vision is a perfect illustration of the internalization of changes in time and space that train travel caused.

A corollary of this is that once the individual internalizes the stimuli, previous stimuli are hardly recognizable. Schivelbusch explained, "Once the traveler had reorganized his perception so that it became panoramic, he no longer had an eye for the impressions of, say, a coach trip."[70] The previously satisfying stimuli no longer register with the observer who now requires greater stimulation.

Let us now apply the concept of stimulus shield to life in the late twentieth and early twenty-first centuries. Some of the writers I will draw upon have intuited Schivelbusch's idea. The following discussion applies mainly to the computer and the mass media. Craig Brod has observed that workers who use computers sometimes display irritation with people and machines that do not perform quickly enough. Brod perceptively noted that, in such cases, the "worker internalizes the rapid, instant access mode of computer operations."[71]

Mander noted that the mass media (including the computer) have sped up our perceptual and nervous systems.[72] Consequently, many of us are bored with a slower pace of life, time alone to read and think, time to be introspective. Because reality is in the media and the computer, to be alone

69. Simmel, "Metropolis and Mental Life."
70. Schivelbusch, *Railway Journey*, 165.
71. Brod, *Technostress*, 43.
72. Mander, *Absence of the Sacred*, 86.

is to be out of touch, to become fantastic. We cannot temporarily with-draw from the company of others to better prepare to rejoin this company later; to conform, one must be encapsulated in the media. In Schivelbusch's terms, we find it difficult to return to the slower stimuli of reading and thinking. Some researchers have discovered that understimulation can be experienced as stress.[73]

Attention deficit disorder and hyperactivity are only the extreme mani-festations of our internalization of the rapid movement of the images and sounds of the media. Jane Healy has studied both disorders and saw their relationship to heavy television viewing and prolonged playing of video and computer games.[74] Attention deficit disorder is independent of hyperactivity, but they sometimes occur together. Both have been diagnosed as forms of learning disability. Attention deficit disorder entails the inability to maintain attention and concentrate on a specific task. Hyperactivity applies to those who cannot concentrate and also have trouble sitting still; their bodies must be in constant motion. Healy cited research on the brain and central nervous system suggesting that overexposure to the mass media and computer may prevent synapses from forming properly in the left hemisphere of the brain so that the ability to read, write, and think is stunted. Someone who lacks facility in language does not understand much of what goes on in school and experiences wandering attention. This mimics the visual media, which moves rapidly from one image to another, from one topic to another.

The ability to read, write, and think is generally declining.[75] The prob-lem is not merely too much television or too much video and computer games but the low quality of conversation and what is read. As Healy noted, students are not expected often enough to read complex sentences and struggle with complex thoughts.[76] Moreover, there has been a dramatic de-cline in the level of discourse in the media, textbooks, and the books most adults read.[77] Given the fast pace of life in a technological society and the rapid movement of images in the media and video and computer games, it is understandable that we have all become mildly hyperactive and suffer from attention deficit disorder to some extent. Hyperactivity and attention deficit disorder are this century's stimulus shields.

The impact of the media (including the computer) includes a loss of memory, personal thought, and feeling. Memory is an integral part of one's

73. Bradley, *Psychosocial Work Environment*, 86.
74. Healy, *Endangered Minds*.
75. Stivers, *Technology as Magic*.
76. Healy, *Endangered Minds*.
77. Stivers, *Technology as Magic*.

existence; it is a product of one's relationship to reality defined by culture and placed into the context of one's entire life. Memory tends to be centered on key events and people that have acquired a certain symbolic quality. There is neither past nor future in the media, only the present. The media aestheticizes reality and turns life into a spectacle to be enjoyed now. There is no need for memory, at least long-term memory, when one lives exclusively in and for the moment.[78]

Visual images have a largely emotional impact upon us.[79] When we watch television, the activity in the left hemisphere of the brain is about what it is when we are asleep.[80] Today, adults as well as children spend much more time watching the media and playing video and computer games than they do reading. What reading they do tends not to be serious.[81] The result is catastrophic. Healy observed that "professionals report that children in classrooms seem to be thinking and learning in increasingly more nonsequential and visual ways."[82] Consequently, there is a decline in logical thought and critical reflection. The image-oriented individual is more readily influenced by propaganda, advertising, and public relations. Healy reported that the use of music with a heavy beat "blocks the capacity for thinking." Children may be learning "to swaddle their brains in sensation-dulling music as an escape from excesses of stimulation in everyday life."[83] The music-addicted mind functions as a stimulus shield.

Finally, Brod discussed the anxiety the computer creates for its users. They tend to make one of two responses: a "struggle to accept the computer" or "overidentification" with it.[84] Those who embrace the computer and make it the center of their lives suffer the "loss of the capacity to feel and to relate to others."[85] In other words, they become like the computer.

The mass media and the computer have led to a loss of memory, thought, and feeling. We are becoming more like machines, which possess none of these capabilities. Our stimulus shield allows us to internalize certain machines—the media and the computer.

78. Ellul, *Humiliation of the Word.*

79. Gombrich, "Visual Image."

80. Mander, *Elimination of Television*, 205.

81. Stivers, *Technology as Magic*, 44–48.

82. Healy, *Endangered Minds*, 132.

83. Healy, *Endangered Minds*, 175.

84. Brod, *Technostress*, 16–18.

85. Brod, *Technostress*, 18.

The Theory of the Technological Personality

Previously, we saw that the technological personality—the aggressively cheerful, garrulous, mildly hyperactive, outer personality—compensated for the lonely, fearful inner personality. But the technological personality serves other purposes as well. The technological personality with its hyperactivity and loss of attention, memory, thought, and feeling internalizes both the tempo and synchronicity of a technological civilization and the machine itself. We engage in frantic activity and communication without much thought or feeling. As Ellul succinctly put it, in a technological civilization, reflex supplants reflection.[86]

All this frantic activity and communication serve to distract the technological personality from the loneliness and fear that the inner personality experiences. Yet these distractions are only superficially and temporarily successful, for the technological personality reinforces the experience of loneliness. The stress that technology directly creates is qualitatively different from the kind it generates by indirectly making human relationships more abstract, impersonal, vague, and competitive. The technological personality is not a stimulus shield to loneliness. All of us differ to the extent that our personalities resemble the technological personality. The individual whose personality is close to the technological personality simultaneously experiences less technological stress and more loneliness.

One of Ellul's great insights is that in a technological civilization, everything is an imitation of technology or a compensation for its impact or both.[87] The technological personality perfectly illustrates this. As stimulus shield, the technological personality *imitates* technology by internalizing it. The technological personality, in turn, *compensates* for the inner, lonely personality (an indirect result of technology) by extroverted cheerfulness.

86. Ellul, *Technological Society*.
87. Ellul, *Perspectives on Our Age*, 48.

Chapter 4

No Place for Men

s it possible to write about the plight of men today without being perceived as partisan? I doubt it, but I will make an attempt anyway. Properly understood, my historical and sociological argument does not "blame" any group—adults, women, fathers, and so forth. The important factors behind the lengthening period of immaturity of men transcends the actions of any group. The problem goes back 200 years or so. Consequently, there is no short-term remedy. Indeed, social problems, which are moral problems as well, defy solutions. The idea that there is a solution to a social problem is part of a technological mind-set. But moral problems are not technical problems. Still we need to understand the issue to be able to face it realistically. A moral response necessitates a realistic understanding of an issue, whereas a technological "solution" treats an issue abstractly in removing it from its socio-historical context.

Much has been said already about male immaturity, about how long it takes boys to become men, if at all. Some boys reject or postpone indefinitely career, marriage, and family, preferring the freedom of a permanent adolescence. Critics and defenders of men, feminists and male role experts alike, are in agreement on the growing prevalence of a lengthening male immaturity.

Girls do better in school than boys, and the percentage of women in college approximates 60 percent. Women now make up the majority of admissions to law and medical schools. American men increasingly eschew majors in mathematics, science, and engineering, once considered majors for men. Traditional male jobs in construction and in factories have been declining as a percentage of the overall job market. Media representations of men include comedic depictions of them as boys who refuse to grow up. Puerile male humor about bodily sounds and functions, excessive drinking, and stupid antics is widespread. In the opposite direction but still stereotyped is the portrayal of men as sexual predators. Men appear to be in retreat, confused about what it means to be a man.

There are numerous articles and books on the subject, many of which are insightful as far as they go. The emphasis is on historically specific factors, including the Vietnam War, the feminist movement, and of loss of jobs that once went to men almost exclusively. My emphasis is on long-term structural changes within which such historical events occur. The advent of a technological civilization (including a media culture), the loss of a common morality, and the politization of group relations are paramount factors.

Historical Causes of a Lengthening Immaturity

This section of the essay refers to both male and female immaturity, the remainder about male immaturity in particular. The most helpful work in this regard is J. H. van den Berg's *The Changing Nature of Man*.[1] As a historical psychologist he situates the emergence of an emotional and intellectual immaturity in the eighteenth century. The period of immaturity has been lengthening ever since.

Beginning in the eighteenth century science and technology begin to replace religion in the cultural establishment of truth. Science is perceived as objective because the scientific method purports to be an objective process that confers objective status on its findings. Science provides us with truth, but only insofar as truth is equated exclusively with facts and their explanation. Owen Barfield goes so far as to say that today technology determines truth.[2] In a radically immanent world in which everything is matter, as applied science technology allows us to manipulate the world to our advantage. Truth is power over reality. Up until this time truth was related to a set of absolute values that religion established; these values were perceived as objective. Truth involved the congruence of human actions and words with these absolute values.

With the ascendancy of science and technology, religion begins to be perceived as subjective, something individuals chose for the sake of their families. The subjectivity of religion appeared to fit the proliferation of religious sects. Religion, then, was a choice one made for family, business, security, health, happiness, or success. If religion was a subjective choice and thus relative, so too were its values and morality. Although it took longer for the latter to sink into the collective conscience. As van den Berg

1. van den Berg, *Changing Nature of Man*.
2. Barfield, *Saving the Appearances*.

observes, God had disappeared, leaving a self-contained material world to its own devices.[3]

Homogenization was a consequence of the new world view. Descartes maintained that matter could be reduced to extension in space. As such everything becomes equal in terms of its composition—matter. Everything material can be measured. Qualities, however, cannot be measured. Qualities depend upon context for meaning, historical, cultural, and interpersonal contexts. Discourse in natural languages allows us to communicate differences and similarities in meaning by supplying the context of meaning. Take for example two young people discussing whether their parents loved them or not. What each person has to do in a successful dialogue is to convey what she means by parental love (which exists in historical, cultural, and a myriad of interpersonal contexts). Even at best no two people perfectly understand each other, because the meaning of words cannot be made specific to one's unique circumstance; otherwise it would be a private language. Discourse is imprecise, but measurement is precise.

With all of life coming under technological control, everything must be measured. Love, for instance, can be measured, if it is a form of matter, differing only in amount. Qualitative judgments are inappropriately turned into quantifiable categories that allow for measurable outcomes (both for individuals and groups). Homogenization suggests a norm of equality that contradicts the inequalities (real differences) of everyday life. For instance, many suggest that women and men should receive the same pay and equal promotions as measured in the aggregate by outcome not opportunity. Because outcome often is an indicator of historical discrimination, equal aggregate outcome becomes a norm that overtakes equality of opportunity. The norm negates individual differences. In some companies women as a group should receive higher pay and be more represented in management than men because more individual women achieved these outcomes than did individual men. The opposite is possible as well. The problem is that historical discrimination, which has an impact on the present, is so intertwined with equality of opportunity that the only apparent way to measure equality is in terms of aggregate outcome. Consequently, we are moving toward ever greater homogenization. Increasingly the individual is being swallowed by the group. Individual differences struggle for existence. The quantification of quality is accompanied by the homogenization of individuals. The contradiction between the norm of equality and the inequalities of everyday life is

3. Barfield, *Saving the Appearances*, 189–90.

both cause and effect of anomie (a term Emile Durkheim[4] made famous) or normlessness. The moral unity of society begins to fall apart.

The loss of moral unity has two far-reaching consequences. The first is what van den Berg terms "multivalency." Multivalency suggests that there is no common meaning in any of our words, actions, and institutions. Each group, association, and organization has its own set of meanings and values. In turn each individual has his own. The cultural context of meaning is destroyed, and no group, association, organization, or individual is able to create and sustain a full culture for itself, because it has to interact with other groups, associations, organizations, and individuals. Hence culture becomes fragmented, ephemeral, and random.

The disappearance of a common morality makes ambiguous relations between statuses in society. Young and old, female and male, lower and upper class, have no obligations to one another because culture provides no guidance. The individual is set loose to fend for herself in a sea of ambiguity. (We will discuss pluralism at length in the next section.)

The second consequence is a discontinuity between generations. Now no traditional values are passed on to the next generation. Old and young, parent and child, begin to inhabit separate worlds; misunderstandings abound. The authority of culture is supplanted by the authority of abstract knowledge. Homogenization produces a radically different type of continuity than a traditional culture did.

A traditional culture did not allow one to predict the future. The future had to be imagined and seized. The reason is that even though the values are traditional their meaning varies as they are applied to new circumstances. As a result the future is open and the present is different from the past. But the values survive from generation to generation.

By contrast homogeneity means that past, present, and future are the same, differing only in quantity—progress. The future is to be predicted and controlled by science and technology. Information supplants values. One must make choices based on the best information. Specialization makes occupational choices difficult for the various occupations and professions, which have become almost invisible to everyday experience. Hence the necessity for vocational counselors. Marriage too is invisible because parents and children live in almost mutually exclusive worlds. Media representations of marriage are at best a shallow and misleading attempt to make marriage visible.

Today, then, children face two obstacles to maturity: multivalency and invisible work and marriage. To become an adult and assume its

4. Durkheim, *Suicide*.

responsibilities, one must now make a "correct" decision based on abstract information. This can take a long time, however, and be fraught with anxiety about the future. Therefore, some postpone decisions about maturity, preferring to remain an adolescent or child indefinitely.

Parents often do not help either. They keep a child a child as long as possible because they too are anxious about their child's future. In traditional societies children quickly became adults in part because they mingled freely in the adult world. Nothing was invisible to them and they shared common values. The moral unity of the entire society included the relations between different status groups. It is no coincidence that childhood and adolescence as distinct stages of life began in the late eighteenth and early nineteenth centuries.

Without the authority of culture, parents must rely on the psychological control of children. Parents cannot make effective use of the authority of information before children are mature enough to appreciate information. This is why techniques of parenting are ineffective unless they include tips on psychological manipulation. Psychological control is a dangerous game, however. It can lead to the absorption of the personality of the child, creating children who are overly dependent on their parents. Van den Berg maintains that parents have a difficult time finding a balance between giving a child too much independence and making a child too dependent. It is not surprising that what we call good parenting today is being overly solicitous. Helicopter parents are only an extreme example of a widespread phenomenon. Made too dependent, children have trouble growing up. A lengthening time of immaturity has continued uninterruptedly since the nineteenth century. Age at marriage has been increasing as has the time to select an occupation. Young men and women live with their parents into their mid-twenties and thirties to a greater degree than in the recent past.

Politization of the Relationship between the Sexes

The decline of moral unity not only leaves parents and children in different worlds, but women and men as well. Lawrence Friedman refers to this pluralism as the "horizontal society."[5] Every group becomes a special-interest group and has no culturally defined relationship to other groups.

Before discussing the horizontal society in greater detail, we must first examine its opposite—a hierarchical society. No one, I think, has done a better job of analyzing hierarchy than anthropologist Louis Dumont.[6] He

5. Friedman, *Horizontal Society.*
6. Dumont, *Homo Hierarchicus.*

argues that modern societies can be characterized by the values of equality and individualism, whereas traditional societies by the values of hierarchy and holism. As Dumont notes, there is no true social order without cultural authority; the latter necessarily entails hierarchy. Hierarchy goes with holism in the following way. Some statuses are higher than others, e.g., old is higher than young, male is higher than female (sometimes it is the reverse). When the hierarchy works well the status difference is small and the difference in power even smaller. What holds the hierarchy together is holism. That is, a sense of the whole, the community, takes precedence over and mitigates differences in status and power. One is first a member of the community and only then an old or young person, a man or a woman. Moreover, holism is based on the idea of complementarity: it takes both old and young, male and female, to make a community. The fundamental division of labor in traditional societies is based on complementarity. There is invariably a tension between hierarchy and holism so that differences in status and power threaten to diminish holism. Power and status can become ends in themselves at the expense of community.

Modern societies value equality and individualism. Observers like Alexis de Tocqueville noticed in the nineteenth century that equality had become the primary value in the United States and that equality went hand in hand with individualism.[7] Equality negates cultural authority. Freed from the control of parents and community, individuals had to make decisions for themselves. Tocqueville found that this individualism resulted in "psychological weakness," a tacit fear of others because relationships without the order of hierarchy had become competitive and uncertain. Individuals now turned to public opinion and government to help them make decisions and to protect them. Consequently, individualism led to collectivism, what Tocqueville termed the "tyranny of the majority." This new collectivism has little in common with the holism of traditional societies.

A horizontal society, Friedman maintains, is one in which "Each status, each identity, has become, as it were, a type of nation, community, or tribe."[8] In a hierarchical society, relations between status groups are normative and complementary. In the horizontal society, by contrast, the relations are anomic and competitive.

We have now become a nation of special-interest groups with vast memberships. With online communication these groups become crowds. Elias Canetti demonstrates that the "open crowd" is about power, growth,

7. Tocqueville, *Democracy in America.*
8. Friedman, *Horizontal Society,* 225.

and equality.[9] Crowd equality is total homogeneity, the divestment of all differences. Crowd equality is transitory, for when one leaves the crowd, inequality reappears. The crowd as well is a hedge against loneliness but only as long as one is immersed in crowd activity. As the crowd grows, its power increases. Today no one dares argue with large numbers of anything, whether people or data.

A paradox of the horizontal society is at the same moment the individual is submerged into the group, the group is advocating individual rights. Individualism results in tribalism, Friedman observes, but tribalism furthers individual interests.[10] As critics from Tocqueville to the present have observed, extreme individualism gives rise to extreme collectivism in the form of public opinion, special-interest groups, public and private bureaucracies, and the media. In desiring power for our group, we are asserting our individual rights, but an individual defined by the amalgam of one's group identities and patterns of consumption. We are members of what Daniel Boorstin calls "consumption communities."[11]

Today we are faced with a spiral of competition for power between special-interest groups. Our concern here is that between women and men. The competition has to be partially masked, for we have the need to believe that the relations between the sexes can be made normative. But reality says otherwise.

Understanding the relations between the sexes means examining the contradictions of power and equality. The exercise of power invariably leads to its abuse. Countless critics have discussed how power corrupts the holder of power—the greater the power, the greater the corruption. Moreover, Max Weber observed that the exercise of power always exceeds the limits of its cultural authority.[12] Consequently men have abused their authority over women. Some would say that men's traditional cultural authority over women is itself an abuse of power. But even if one views the cultural authority of men over women as legitimate in certain circumstances and for certain activities, the abuse of power is inevitable. In Dumont's concept of hierarchy, men and women have complementary statuses that while they differ in power can be held together to a greater or lesser extent by holism. Yet authority necessarily implies some degree of inequality.

In *The Feminization of American Culture*, Ann Douglas convincingly argues that the status of women was worse under industrialized capitalism

9. Canetti, *Crowds and Power*.

10. Friedman, *Horizontal Society*.

11. Boorstin, *Americans*.

12. For a discussion of this point, see Ricoeur, *Lectures on Ideology*.

in the nineteenth century than it had been in agricultural society.[13] Previously there had been a division of labor between women and men, with each sex having authority in its domain. Now the middle class husband had full authority over his wife and children. He was perceived to be God's representative on earth. Stay-at-home wives, Douglas maintains, were in charge of emotional sustenance to the family and expected to subordinate their lives to those of their husbands and children so much so that they retained no independent identity. Men had almost unlimited power. From the nineteenth century onward the feminist movement criticized and resisted the abuse of power by men, including at times any inequality of power between the sexes.

As Tocqueville observed, in a time of equality we criticize those in power but love power itself.[14] Not only love the power but envy the powerful. Søren Kierkegaard observed that in a time when public opinion, which is based on desire and fear, is supreme, envy is the negative unifying principle of society.[15] Consequently, we both resent and envy the powerful. Tocqueville further noted that the more equal the circumstances of life become, the more intolerable the still remaining inequalities appear to be.[16] We want equality, perfect equality, now. He did not say this so much as to criticize those who seek equality, but more as an observation about human nature.

So what exactly is an equality of power? It is a chimera. Dostoyevsky provides us with a meditation on power in *Notes From Underground*.[17] He exposes the master/slave logic of the modern world. The "underground man" lives in a world with three levels: superior, inferior, and equal. The individual or group whose goal is equality proves by this that it is inferior. Competition is only pleasurable when it is between equals. As soon as one party wins, the sense of superiority wanes because the loser is not equal anymore. Competition, then, rules out equality because everyone becomes a master (superior, winner) or a slave (inferior, loser). The inferior is bitter and envious, whereas the superior is left to savor a hollow victory. By itself the competition for power, even for an equality of power, is selfish and destructive. Only an equality based on respect, cooperation, and unselfish concern for the other can result in true equality. With the dissolution of

---

13. Douglas, *Feminization of American Culture*.

14. Tocqueville, *Democracy in America*, 671–74.

15. Kierkegaard, *Present Age*.

16. Furet, "Conceptual System," 187–88.

17. This interpretation of *Notes From Underground* is taken from Tzvetan Todorov, "Notes From Underground."

moral unity, all relations involve a competition for power with equality always in the future.

Men have abused their power and lost any semblance of moral stature. What Dostoyevsky could not have anticipated is victims of power—the powerless or less powerful—now have high moral standing. Women can criticize and ridicule men in public discourse, but the reverse has become next to impossible. We have institutionalized moral outrage. In the competition for power one must never admit that things have become more or less equal, for then one would lose one's moral superiority. The winner in the competition for power has no moral standing. Those who desire greater power, however, believe that they will not be corrupted by power. Power not morality is the goal of the competitors. Morality is the basis for criticism of the powerful, but not the goal of the powerless.

Tzvetan Todorov reminds us that the love of one's group as with xenophilia is a form of prejudice in that it makes all members of the group good simply by being part of the group and at least covertly touts the superiority of one's own group.[18] Xenophobia is the just the opposite—a hatred of a group all of whose members are considered to be despicable and which is therefore inferior. The two go together in practice: One loves one's own group at the same time one hates the other group, one's competitor. In this logic, men are bad, women are good. If becoming a man means becoming something bad, some boys may resist the urge to grow up.

The Feminization of Work

Traditional male jobs in farming and industry have been declining as a percentage of available work. Jobs in the service sector now comprise over 80 percent of overall jobs. Robotics has usurped many factory and construction jobs, and artificial intelligence is gradually replacing intellectual labor. In many cultures women are expected to perform what Arlie Hochschild[19] calls "emotional labor." This is related of course to the woman's culturally defined role of nurturer. Many occupations in the service sector largely involve emotional labor, such as teacher, social worker, and counselor. A key component of emotional labor is the ability to manipulate one's feelings and personality so as to have a positive impact on others. This involves the suppression of genuine feeling in the interest of appearing friendly and agreeable. The goal is to turn the customer into a return customer who rates the service highly.

18. Todorov, "Contacts among Cultures."
19. Hochschild, *Managed Heart*.

What Hochschild and others have been describing is part of a larger phenomenon in business and government—the role of personality in work. In the 1950s David Riesman[20] and William Whyte[21] among others described how important personality and emotional adjustment had become as a part of work. Riesman argued that the main product today is personality and that we have become consumers of personality. An other-directed personality had become essential in business and government. The other-directed person had an extroverted personality whose main goal was to be popular in the peer group. At the same time one adjusts one's viewpoint to fit the expectations of others. As Whyte understood, the "organization man" was required to fit in, to be accepted and liked by co-workers. The peer group swallows up individuality. In a more recent study of corporations, Robert Jackall discovered that a pleasant personality counted more than skill (as long as one was minimally competent) in obtaining promotions.[22]

One technique of management is the human relations approach used by almost all organizations as a primary or secondary strategy for motivating employees. This approach attempts to create a "family" atmosphere in which employees identify with the organization. The peer group (one's family) is used to maintain morale and increase productivity. The goal is to turn the employee into a "team player," that is, one whose adjustment to the organization is total.[23]

By contrast, in the nineteenth and early twentieth centuries, male entrepreneurs were "rugged individuals," plain-spoken and uninterested in manipulating the emotions of employees. They gave orders and expected them to be followed—they were independent of the feelings of others.

Because women have often been placed in a dependent position, they needed to be sensitive to the feelings of those who had authority over them. It was a survival strategy as well as a way of gaining some leverage in the relationship. It now appears that almost everyone at work is expected to play the role of the subservient employee. Work has been "feminized." Is it any wonder that some boys find work at odds with what they thought masculinity entailed?

20. Riesman, *Lonely Crowd.*
21. Whyte, *Organization Man.*
22. Jackall, *Moral Mazes.*
23. Stivers, *Technology as Magic.*

Male Compensation

Masculinity is under attack: male identity is ambiguous with charges of abuse of power widespread and with the decline of men's work. There are ways to compensate for a disappearing masculinity, however. These compensations include comics and graphic novels, video games, and paramilitary activities. I will briefly discuss the latter two activities because they engage the player both physically and mentally.

One hundred and fifty-five million Americans play video games and spend over twenty-two billion dollars a year. The single largest group of players is white men in their thirties. As a point of comparison, Americans spend 10.4 billion dollars on movie tickets. Video games emphasize chase, attack, and death. Violence is thus replete in video games whether the target is a zombie, a traitor, or a hardened criminal. And those killed can return from the dead to be killed again. The action is frenetic: speed is everything.

Paramilitary activities range from reading the *Soldier of Fortune* magazine, attending social events with actual soldiers of fortune, to engaging in simulated activities with weapon in hand. Paint ball is the most popular of the latter activities, but a smaller number attend camps where they can play at being a soldier using real weapons. Paramilitary activity is incorporated into video games as well.

James William Gibson suggests that paramilitary activity is an escape from the recent military failures of the United States, the loss of manufacturing jobs, and the movement of women into formerly male domains. He maintains that the paramilitary activities represent an attempt to return to an idealized past when America was dominant militarily and economically and men were in charge.[24]

In *God in the Machine*, Liel Leibovitz argues that video games are centered on spiritual pursuit. The games teach the player his place in the world, to love the designer of the game (a god), and to give up all other ways of being (the secular).[25] Others take video games at face value: an adventure in which one attempts to become a hero.

Paramilitary activities too allow one to become a hero vicariously. Rambo is the perfect hero, for he takes on everyone—the obviously evil and the covertly evil, the establishment. Video games and paramilitary activities promote heroism, every man's fantasy.

The idea of a hero is historically and culturally specific. J. Huizinga discusses how court life and a military career provided little opportunity for

24. Gibson, *Warrior Dreams*.
25. Leibovitz, *God in the Machine*.

heroism in the late Middle Ages. Consequently chivalry as a cultural form emerged. The tournament and chivalrous adventures as told in story and lived out were compensations.[26]

Today the context of the hero is technological. We live in a civilization in which technology is omnipresent and perceived to be omniscient and omnipotent. The hero today is the weapon. In paramilitary activities, it is a gun, preferably an automatic weapon. In video games the range of weapons is expanded to include even magic. Technology, however, is the stepchild of magic, as Marcel Mauss has demonstrated.[27] Both are about efficacy. Technology is so powerful today that we are beholden to it. Hence the hero is a technological device to which its user is subordinate. Moreover, one infers goodness or heroism from the power of the weapon. David Riesman argued that what makes a superhero good was that she was victorious.[28] The more powerful, the winner, is good. From power comes goodness. We have thus made power a value.

Paramilitary activities and video games have helped make the hero ambiguous. The hero often takes the law into his own hands, answering only to his own understanding of right and wrong, no matter how self-serving it may prove. Such compensatory activities can readily provide a male adolescent with a negative identity—a hero who violates all standards but his own. Erik Erikson has described how an individual, especially an adolescent, may choose a negative identity when no positive identity seems achievable.[29] The hero who is subservient to his technological device is a negative hero out of control.

Video games and paramilitary activities are a form of play. Play becomes corrupted, Roger Caillois tells us, when it is incorporated into everyday life.[30] When this happens the escape from everyday life—play—becomes an obligation, a compulsion, a passion. Today we readily talk about addiction to video games and gun ownership and use.

J. Huizinga furthers this argument by describing the puerilism of the modern world—when play becomes work and work becomes play. In other words, we take play and games seriously and regard work and politics as only a game. While we have scant control over the economy and polity but as participants and fans, we have a more direct and immediate influence on the outcome of a game. Responsibility for the important decisions in life,

26. Huizinga, *Waning of the Middle Ages*, 67–107.

27. His view of magic is discussed in Stivers, *Technology as Magic*.

28. Riesman, *Lonely Crowd*.

29. Erikson, *Identity*.

30. Caillois, *Man, Play and Games*.

moreover, rest in the province of the technology, organizations, and experts. Our responsibility lies in accessing an expert or organization to deliver the desired service. We have outsourced responsibility. Play and games, however, can provide responsibility to the player and fan.[31]

Comics and graphic novels, video games, and paramilitary activities allow the adolescent male (whatever his age) to remain in a suspended state of animation, to remain frozen in place. They are compensations but offer no way out to achieve manhood.

Visual Stereotypes of Men

Male immaturity has not escaped the media.[32] Television and movies, as if to correct an idealization of men in the past, tend to represent men in two stereotyped ways—as predator and as boy. The former is tragic, the latter comic. Sitcoms like *Two and a Half Men* and *Seinfeld,* and movies such as *Boy, Animal House, Jack-Ass, Old School,* and *Step Brothers* portray the comical side of men who are still boys. There has been an explosion of movies about the antics of immature men that feature excessive drinking, sports fanaticism, gross bodily noises, random sex, and stupid actions. In these movies men are shown to be terminally stupid, as in *Dumb and Dumber.* Not that girls have not been pulled into these shallow representations, especially those involving sorority houses.

Concurrently men are portrayed as predators, as those who emotionally, physically, and sexually abuse women. Countless television programs and movies reveal women as victims of male power. The recent television series *Big Little Lies* dramatizes how a group of women avenged the rape and physical abuse one man inflicted on two of them. The movie *Room* shows a sadistic man terrorizing a mother and her son. *Two and a Half Men* combines the two stereotypes by depicting a sexual predator comically.

Literature, movies, and television can either represent characters realistically or idealistically. In either instance, the danger of playing into or reinforcing an extant stereotype is always present. Stereotyping involves treating an individual exclusively as a member of a social category and then generalizing about that category in either negative or positive terms. We tend to think about stereotypes as negative representations, but they can be "positive" as well. In truth all stereotypes are negative in destroying individuality. Sometimes, however, the positive stereotype is used to counter the

31. Huizinga, *Shadow of Tomorrow,* ch. XVI.
32. Cross, *Men to Boys.*

negative stereotype. Some have observed that the sitcom *The Cosby Show* attempted to create a positive stereotype to offset prevalent negative stereotypes of black families. Visual images in the media create stereotypes even when that is not their purpose.

The overwhelming dominance of artificial visual images in the media (visual images we create with our technology) has led to a great reversal in which discourse is now subordinate to visual images. A technological culture is visual, and the images in the media are technology's language, so to speak. Technology manipulates our material existence and attempts to make material that which is not through quantification and visualization. Words are "explained" by visual images.

Discourse has been moving in opposite directions simultaneously. Because of advertising, public relations, and propaganda, many words have become vague and retain only an emotional meaning. The word *democracy* can refer to almost any political regime; the word *revolution* to any innovation, e.g., a revolution in kitchen cleansers. Either the vague word has too many referents, or it does not refer to anything that exists. But along comes the visual image to provide us with an operational indicator of the word. Now love "means" a visualized hug or kiss. This produces concrete thinking in which we associate image with image, instead of word with word.[33]

Visual images have become "norms in a world without meaning," writes Jacques Ellul.[34] By this he means what was previously described by van den Berg as multivalency. Visual images in the media are compact stereotypes, and the more we receive information dramatized by the media, the more we think in stereotypes. The visual stereotypes of man as predator and man as boy harm the individuality of each young man. The stereotypes offer negative identities at a time in life when identity formation is most poignant.

Edgar Friedenberg maintains that fewer adults today have passed through adolescence and become mature adults than in the past. To become mature, he maintains, one needs to become emotionally independent from one's parents and teachers, assume responsibility for one's actions, clarify one's experiences, and think for oneself. Parents and teachers who make children overly dependent and preach adjustment and conformity to them thwart the maturation process.[35] If boys (and girls) have relatively few adult role models in life but only stereotypes of men and women in the media, we should not expect many to reach psychological maturity.

33. Ellul, *Humiliation of the Word*; Stivers, *Technology as Magic*.
34. Ellul, *Humiliation of the Word*, 147.
35. Friedenberg, *Vanishing Adolescent*.

No Place for Men

Male power is under attack and by association masculinity itself. The attempts to redefine masculinity have been failures. Androgyny never caught on; it was based on a contradiction. One cannot simply take the attributes of women and men we like best and cobble together a single sexual identity. In the real world the best characteristics of women and men, such as kindness and sternness, only exist as complementary virtues. Each sex, then, possesses certain attributes in relation to its opposite sex, who embodies a different but complementary set of characteristics. For instance, a woman might embody kindness, nurturance, and caring, whereas the man might represent the qualities of discipline and self-control. If men and women exist in a moral relationship of marriage or deep friendship, each can acquire the other's attributes in a secondary way. Some traditional societies believe, at least tacitly, that it takes both male and female to make a complete human. If androgyny were to occur (and homogenization may yet make it a reality), the complementary virtues of each sex would become vague. It is only by acting in relation to one's opposite that one's own attributes are established, clarified, and developed.

The attempt to make men more like women, "male feminism," is likewise making few inroads if only because some women have so readily abandoned female characteristics in an attempt to act like men in business and politics. The implication is that the traditional female role is less important than the male role (protests to the contrary). A greater equality of the sexes could be achieved if we regarded each role in the division of labor as equally valuable (thereby eliminating a single hierarchy of authority) and made the division of labor highly flexible. In traditional societies, out of necessity women sometimes had to assume the male role and men the female role.

The world of men is being opened to women, and that world is shrinking for men. Women add the world of men to what constitutes being a woman; men have to share this world because they have monopolized and abused their power. Women have gained a world, while men have lost a world without any clear alternative. Vilified and disgraced, men have no place in society. The way to bring men back into society—reconciliation, love, and forgiveness—likewise has no place in a society that understands human relations as relations of power.

Chapter 5 _____

The Computer and Education: Choosing the Least Powerful Means of Instruction

Originally published in the *Bulletin of Science, Technology and Society* 19, no. 2 (1999) 99–104.

The computer lies beyond serious criticism. Technological fundamentalists (Wendell Berry's term) demand that everyone use the computer for as many intellectual tasks as possible and view it moreover as a panacea for education's many ills. Moderate technophiles do not see the computer replacing the teacher or making her subordinate to it, but they welcome the advancement of educational computing as much as their fundamentalist allies. The moderates want to have it both ways: the continued improvement of the computer and, as a result, the progress of human intelligence and judgment. They are the greater threat; for in claiming that the computer is now and will remain under full human control, they lull us into a technological slumber. Those of us who raise serious criticisms about the computer are referred to as techniphobics (those who have an undue fear of the computer), but one should read that as technological heretics (those who do not believe in the truth of the computer).

In the following essay, I wish to make the strongest case possible against the use of the computer in education. My starting point is Jacques Ellul's insight that the computer and television are the chief sources of a nonpolitical totalitarianism in advanced societies.[1] I will attempt to elucidate Ellul's thesis, at least in regard to the computer, by examining the consequences of its use in education: (a) its effect on the language, logic, and content of what is learned; (b) its tacit worldview that effectively teaches students a cynical lesson about life; (c) its part in the fragmentation of the personality of the student; and (d) its attack on the moral character and

1. Ellul, *Technological Bluff*, 384–400.

freedom of the student. In short, the computer instructs students about life in ways teachers never intended, and it molds them more completely than the most authoritarian teacher ever imagined possible.

I will spend less time on the first point because others, such as Stephen Talbott, have written about it extensively.[2] The computer necessarily reduces discourse and the words that comprise it to their most abstract meaning, it eliminates dialectical thought, and it reduces knowledge to information (unless it simply reproduces a text). The computer is the slave of speed and efficiency, but human language is situated within the slow movement of history. Paul Ricoeur has argued that the sentence (metaphor is included here) is the smallest unit of meaning; that is, the meaning of language lies in its use.[3] Words acquire their meaning from the context of their use. The dictionary definition is only the most general or abstract meaning of a word. But except for purely technical terms, most words have a number of meanings; their meaning is ambiguous. The use of the words in discourse supplies a context and thus reduces some of the natural ambiguity of words. If words had only one precise meaning, then they could not be used in a variety of contexts and could not change. But we know that all natural languages evolve.

Take the word *love*, for example. The meaning of *love* varies according to the culture and historical period, according to one's own experiences, and according to the situation to which one is applying the concept. In the late middle ages in the West, to love a child meant to discipline a child (in the sense of building moral character).[4] This was not done for the most part without some expression of affection. But the affection was subordinate to ethical considerations. Today, we tend to place discipline and affection into distinct, if not mutually exclusive, categories so that love is often identified with an affection that is sentimental (expressed as an end in itself).

Qualitative concepts such as love are paradigmatic; within their appropriate cultural setting—for example, familial—they serve as examples for us to apply to our own past experiences in the family and to new family situations, whether our own or those of others. Qualitative concepts such as love, as they are applied and lived out, entail numerous imaginative judgments, both conscious and tacit, that a computer that follows rules could never make. Real education involves helping a student understand, appreciate, and apply such qualitative concepts to a variety of contexts.

2. Talbott, *Future Does Not Compute.*
3. Ricoeur, *Interpretation Theory.*
4. Gottlieb, *Family in the Western World.*

Over against this educational ideal, the computer deletes a culture of experience and symbolic meaning and, for it, substitutes a monolithic culture of abstractions. Whether in artificial intelligence, expert systems, or educational software programs, words retain one meaning—the most general and abstract. The computer turns all words into technical terms. But as Owen Barfield has demonstrated so well, abstract terms are more or less meaningless; they function as things.[5]

Ellul forecast that the logic of the computer will lead to the death of dialectical logic.[6] The Boolean logic of the computer, like all formal logics, eliminates contradiction. However, the logic of existence is dialectical. All formal logics, Kierkegaard noted, assume the identity of thought with existence, but reality reveals their separation.[7] That is, the concept does not begin to exhaust the meaning of that to which it refers. Not only is there a dialectical relationship between thought and existence but also existence itself, both at the level of the individual and of society, is fraught with contradiction. Metaphor, which is the origin of all qualitative concepts, is an expression of the ambiguity of life.[8] Metaphor does not claim that "love is a rose" but that "love both *is* and *is not* a rose"; in other words, it "is *like* a rose." Similarity is not identity. For all the claptrap today about critical thinking, few of its exponents seem to realize that critical thinking is dialectical thinking. For many, it would seem only to be logical thought applied to an issue in the real world. However, it is only dialectical thought that allows us to be free, as Kierkegaard in the nineteenth century and Ellul in the twentieth century have demonstrated in their writings and in their lives.

How does the computer affect the content of what is taught? In analyzing several educational simulation programs such as *Oregon Trail* and *Jenny of the Prairie*, C. A. Bowers notes that educational programs tend to emphasize the actors in the story as "autonomous decision makers,"[9] rational and data driven (not unlike computers), at the expense of the tacit cultural context within which decisions are made. Bowers attributes this in part to the authors of the software and in part to the logic of the computer. It appears that the authors resemble their tool. The computer completes what textbooks in the humanities and social sciences have already begun: taking knowledge out of its fuller cultural context and thus reducing it to information.

5. Barfield, *Poetic Diction*.
6. Ellul, *Technological System*.
7. Kierkegaard, *Concluding Unscientific Postscript*.
8. Wheelwright, *Metaphor and Reality*.
9. Bowers, *Educational Computing*, 36.

However, the computer teaches students more than epistemological lessons about language, logic, and information. Pretending to be the font of all knowledge, it formulates a tacit worldview for the student. The myth of the computer is that it maximizes human choice about information; the reality is that it establishes totalitarian control over information. The computer is the most powerful force within the technological system.

In the nineteenth century, Karl Marx effectively demonstrated that money had become the most powerful force in Western Europe.[10] He argued that the universality of its being was a sign of its omnipotence. It was the universal indicator of worth: every object, action, and human quality had to be measured in terms of money. It became more evident in the decades after Marx's initial essay when insurance companies and courts of law assigned a monetary value to the human body and even to human affection. He went on to argue that money mediated our relationship with ourselves and with others. Money overturned the natural order of ability so that those with money could appropriate the labor and products of others (a patron without artistic talent purchases the paintings of an artist) and that the talents of those without money would shrivel up. Furthermore, money led those who had nothing in common to enter into contractual arrangements, while causing the natural bonds between people to be broken (money problems can eventuate in divorce, alienation of parents from children, and damaged friendships). Marx concluded that money was a "visible divinity" and an "almighty being." In effect, what Marx was saying was that money had become sacred. People's understanding of something as sacred depends on their tacit perception of it as the ultimate power and reality.[11]

Has the computer (technique in a more general sense) replaced money as the chief sacred of modern societies? Has it become the universal indicator of value? Yes, I think so. As Ellul and others have indicated, the reproduction of value is less dependent on capital today than on technical information.[12] The use of and access to money is less contingent on the creative activity of the entrepreneur than on the organization of technical information to coordinate increasingly complex economic activities.

As technical rationality and information render culture ever more abstract, our relationship to ourselves and others follows suit. Technique and technical information mediate my relationship to my abilities. First, the more powerful the technique or tool, the less necessary human skills become. The more time workers devote to the computer, the more their hands-on skills

10. Marx, *Economic and Philosophic Manuscripts*.
11. Eliade, *Sacred and the Profane*.
12. Ellul, *Technological System*.

deteriorate.[13] It is evident that computer-dominated work is the fate of most workers in the future. Second, large numbers of people can only understand themselves through the technical information of popular psychology, for example, self-help groups, positive thinking, and new age thought. At the same time, access to technical information through the computer permits those who do not have the ability or understanding to perform an almost infinite number of tasks. One can employ a computer program even if one knows next to nothing about the activity in the real world, for example, a program for expert advice on investments can be used by one who has little knowledge or experience in the stock market.

The computer likewise mediates my relationship to others. More and more people communicate with one another through email, an impersonal medium. Someone told me recently that she preferred the anonymity of email discourse because there was little chance of being hurt the way one could be in person-to-person relationships. The computer both damages (makes impersonal) real human relationships and also permits one to communicate with virtually everyone (creates false relationships). A professor at my university was discussing with a former student and me how email is reducing discourse to snippets of information, emotional outbursts, and elliptic sentences. The professor and the graduate student welcomed this development but vaguely did wonder about the future evolution of language.

The computer, the main processor of technical information, is a visible divinity that appears to be omnipotent. Why else would otherwise intelligent people believe that artificial intelligence will someday simulate human intelligence? Because given the infinitely greater capacity of the computer to store information and process it rapidly, artificial intelligence would then surpass human intelligence. Some moderates may say about the computer, "Garbage in, garbage out," but for many people, what comes out of the computer is both reality and truth.[14]

What worldview is implicit in this apparently all-powerful technique? Ellul notes that one of the major consequences of the widespread use of the computer is the creation of an excess of information.[15] Too much information becomes disinformation. The Internet is replete with random and incoherent information. There is some bit of information for every conceivable taste, hobby, and interest. One "surfs the net" in search of the new, the exciting, the fantastic, and the ridiculous. We become connoisseurs of information, most of which is useless for one attempting to lead a meaningful

13. Zuboff, *Age of the Smart Machine*.
14. Ellul, *Humiliation of the Word*.
15. Ellul, *Technological Bluff*.

life. The first thesis of the worldview of the computer is that the world is best understood by the greatest quantity of data possible. Or in other words, understanding equals quantity.

This excess of information eventuates in a "broken vision of the world." One loses a sense of the past, the ability to evaluate the present in terms of the past. The triumph of the computer and the visual images of the media produce a "culture of forgetting."[16] Studies indicate that our memories decline to the extent that everything is done for us (the computer) and shown to us (the images of the media). Memory centers on events that have acquired emotional, intellectual, or moral significance to us. The past understood in either mythological or historical terms acquired meaning through key events that took on symbolic meaning. The information from the computer, as we have previously seen, is abstract, logical, and quantitative, but ultimately meaningless. The cynical worldview of the computer is that we live in a random and meaningless world about which the omnipotent computer can generate an infinite amount of information that we can exploit to our advantage.

What are the consequences of the widespread use of the computer for the student? First, it is a force in the fragmentation of the personality. The computer like all techniques appeals to our will to power; it provides us with a false sense of power: the computer's power is my power. In contradictory fashion, however, it contributes to a "confused sense of impotence."[17] Because that power is really not my power. Moreover, the increase in information about technological means and technological objects may lead to a failure to make a choice. Too many choices and too much information about them can create a situation of sensory and intellectual overload with a resultant lethargy.

Any increase in knowledge and power demands greater self-control of the human using them.[18] But the excessive information of the computer and the media turns us, Ellul notes, into "exclusive consumers," with the result that even shopping for and acquiring consumer goods becomes a means of gathering and using information.[19] That is, we become connoisseurs of products. Even if one only owns an inexpensive and simple stereo system, one can become a connoisseur by gathering sophisticated information about stereo systems. The next step is for the connoisseur to become an addict, one obsessed with a product and the information about it. The tacit goal of all

---

16. Ellul, *Technological Bluff*, 330.
17. Ellul, *Technological Bluff*, 331.
18. Bergson, *Two Sources of Morality*.
19. Ellul, *Technological Bluff*, 330–31.

advertising, which provides much of the information on the Internet, is to turn us into connoisseurs or addicts of as many products and as much information as possible. Then, the computer tends to create addicts with contradictory feelings of omnipotence and impotence. We become secondary and subordinate to the goals of advertising and the consumer objects whose representations envelop us. No self-esteem class can begin to reconstruct the student's personality, a personality that has been fragmented and reified into a desultory collection of consumer identities.

Related to the assault on the integrity of the student's personality is the attack on her moral character and freedom. Information mainly shapes behavior, but qualitative knowledge largely develops character. A moral judgment requires a tacit, practical knowledge of culture and history to inform the experiences that help one evaluate an action, its context, and immediate consequences. However, a technical problem demands detailed, quantitative information about the probability of an outcome and the most efficient way of proceeding. The more powerful the technical knowledge, the less human choice and judgment there is about that knowledge. The qualitative knowledge involved in a moral judgment leaves the individual a certain freedom of imagination in applying a principle or exemplar to an action because of the ambiguity of all qualitative concepts and contexts to which they are applied. No two cases are ever the same, and making a moral judgment involves taking into account the action, the motive, and the consequences. For example, is a parent spanking a child out of anger or because he believes it will benefit the moral development of the child?

The computer specializes in technical information, which may be applied inappropriately to moral situations. Christina Sommers has analyzed how ethics is taught most often in public schools.[20] Some variation of either values clarification or cognitive moral development is used often. Virtues are not taught; rather, students are enjoined to discuss their moral preferences with the assumptions that there is no right answer and that we all have to be tolerant. Within this relativistic framework, there is the vague hope that students will become less individualistic and particularistic in outlook and more communitarian and universalistic. When it comes to action, Sommers notes, attention invariably is directed to political and technological solutions to moral problems so that governmental bodies and corporations become the moral "agents." Following this approach, a teacher can employ the computer to assist students in researching the issue and "choosing" the appropriate organization to solve the problem. For example, how do we best improve the lot of the homeless? With jobs created within the private sector or with

20. Sommers, "Ethics Without Virtue."

an increase in welfare benefits? Moral judgment in this approach is reduced to consumer preference that some organization will act on.

Ellul always argued that moral judgment was the highest expression of human freedom.[21] The distinction between moralism and moral judgment is critical here. Moral judgment can bring about freedom, but only free people can exercise moral judgment. Moralism involves making a moral judgment that one never acts on or lives out in one's life, that involves no risks, and that allows one to feel superior to others. Genuine moral judgment entails putting into practice one's beliefs, risking the disapproval of others, and realizing one's own moral inadequacies; hence, moral judgment makes one responsible for what is judged.

The spoken word is the most appropriate medium for moral judgment. Do you stand behind your words? Do you keep your promises? As we have seen already, the computer tends to deconstruct the meaning of words and, thus, to make abstract such qualitative concepts as freedom, love, and justice; moreover, it permits the anonymous discourse of email, where one can say anything, no matter how preposterous or hurtful, without any risks. The computer encourages the most irresponsible discourse yet known; in effect, it destroys moral character and teaches the student a tacit lesson few of us would support: freedom exists without responsibility.

In this respect, the computer turns life into a game. This point goes beyond playing computer games; it applies to all of the computer functions in which the user participates. Perhaps the Internet is the most dramatic manifestation of this. Life becomes a game of acquiring the most information and the most interesting information on a variety of topics. One shares that information with others as a way of feeling good about oneself. Here is the victory, at once both easy and therapeutic. And after all, what is life that is merely a game if not freedom without responsibility?

Individual freedom is premised as well on some distinction between private and public sectors of life. In the West, this distinction, although quite different in each context, has persisted from Greek and Roman to Judaic-Christian influences. However, totalitarianism blurs this distinction: it makes everything public. The citizen can have no secrets; he must be transparent. Totalitarianism, as we have experienced it in the twentieth century, is coercive and political; it establishes rules by fear. However, the most recent totalitarianism is not political but technological; it is reflected in Huxley's *Brave New World*. The citizens are not controlled by the police but by psychological technicians. Power is exercised through positive reinforcement and mind control; it has become therapeutic.

21. Ellul, *Ethics of Freedom*.

The computer and television, its compensatory ally, have convinced us to make public whatever used to be private. Confession is a major form of discourse: talk shows, the news, and email feature individuals who reveal everything about themselves.[22] There are several reasons for the efface-ment of the private sector. First, reality is on television and the computer. We experience reality through the dominant medium of communication. The visual images of the media and the abstract information of the com-puter have supplanted language as the dominant medium.[23] We are only real when information about us is circulating through these media. Second, the anonymity of discourse on television makes self-revelation painless. Third, all information belongs to everyone, according to the ideology of technol-ogy. To hold back information not only would deprive others of it but also would prevent the computer from becoming omniscient. For technology to run its full course, all possible information must become actual. There can be no secrets, whether about the universe or the individual.

The genius of language is that it allows us to reveal and conceal ourselves at the same time. The ambiguity in meaning of the terms of natural languages means that no one fully reveals herself to another even if she intends to; at the same time, the mystery and freedom of each individual is preserved. With nothing private, there is no freedom and no individuality.

However, the reverse is just as evident: the public has become private. The consumer identities and celebrity identities that television and the computer offer us appear in a condition of cultural meaninglessness to be the exclusive means of acquiring a sense of individuality.[24] Totalitarianism can go further than this: to invert violently and completely the relationship between public and private.

I wish to conclude by reflecting on Milan Kundera's analysis of Franz Kafka's literary writing about bureaucracy. Kafka's major concern is with the functionary, one who simply follows orders. The technocratic universe of the functionary is characterized (a) by obedience to authority, no freedom only rules; (b) by actions that are exclusively technical and thus without meaning; and (c) by an abstract and impersonal relationship to work and to others.[25] The computer has turned us into functionaries because, as the leading part of the technological system, it tends to destroy the meaning of language, reinforces and deepens our abstract relationship to others, and proliferates

22. White, *Tele-Advising.*
23. Ellul, *Humiliation of the Word.*
24. Stivers, *Culture of Cynicism*, 158–59.
25. Kundera, *Art of the Novel.*

technical information that must be obeyed. Technology may be creating a global village, but it is a Babel peopled with functionaries.

Yet, what can we teachers do about the omnipresent computer? I do not think we have the right to decide for students whether they should learn how to use a computer. In a technological civilization, there is no choice about whether or not to use the computer. It is imposed on us as a necessity and an absolute good. However, if teachers respect the freedom of their students, they will use the least powerful means of instruction. They will eschew, as much as they can, audiovisual aids and the computer. These techniques are used to manipulate students into learning or to make learning more efficient by stunting the verbal skills of the student. There are no shortcuts to real learning because it takes place within the slow pace of thinking in and through language. Language is the least powerful means of instruction and the one that best respects the freedom of the student. Because every natural language harbors within itself a possible revolution of new meanings and liberating thoughts.

Finally, we can teach students—commensurate with their age, maturity, and experience—about the detrimental effects of the widespread (there is no other) use of the computer. In so doing, we will be making a moral judgment, taking a risk, and exercising that transitory and elusive virtue we call freedom. We will be teaching by example.

The Need for a "Shadow" University

Originally published in the *Bulletin of Science, Technology and Society* 26, no. 3 (2006) 217–27.

The liberal arts are dying. They no longer make up a living tradition, ensuring some measure of continuity with the past. That is, the liberal arts do not inform the lived experiences and practices of everyday life. Does anyone really believe anymore in the promise of the liberal arts, in the ideal of a well-rounded human being who pursues wisdom and truth as much as power and a good job? More often than not, we justify the liberal arts as thinking skills that are transferable. In some instances, we justify them as pleasurable endeavors, as an end in themselves. But even this latter reason, which I do not reject altogether, was premised to some extent on an appreciation of history, literature, art, philosophy, and language as integrated within a larger culture.

We teachers tend to blame the students for being lazy and unappreciative of higher learning. Concurrently, students often perceive the content of our courses as irrelevant to their lives and future vocations. They are right, of course. J. H. van den Berg, Dutch psychologist and historian, argues that when he and his secondary school classmates, early in the past century, read Shakespeare, Goethe, Cervantes, and Hugo, they understood these writers not because they were more intelligent than students today but because the attitudes, emotions, convictions, conflicts, and decisions expressed by the characters in these narratives were similar to their own.[1] All that has changed. Students, for the most part, do not appreciate the literature, history, art, and philosophy of the past because their experiences, the context of their lives, are radically different. The context of their lives today, and increasingly ours as well, is formed by technology and visual images: the Internet, the shopping mall, television,

1. van den Berg, *Changing Nature of Man.*

the movies, advertising, and video games. The wisdom and experiences of the past are irrelevant in a technological civilization.

The demise of the liberal arts goes well beyond teachers, students, and universities; it is reflective of long-term changes in Western civilization. Until the nineteenth century, the liberal arts had been taught within a normative context, a set of beliefs and assumptions about human nature, the meaning of history, and the order of society, whether classical, Christian, or humanistic. This normative context has been transformed by technology and the universe of visual images so that morality as an open system of symbols and experiential paradigms is being supplanted by a system of technical rules and visual images whose "meaning" is either efficiency or normality. The university, including the liberal arts, is just one institution among many that is being transformed along these lines. A note of warning: I am not opposed to either technology or to visual images. What I am against, and this should become clear later, is technology's domination of life and the corrosive effect the visual images in the media have on discourse. Some of you are going to object that discourse has been a major force in the subjugation of women and other minorities. This is undoubtedly true. Discourse often is ideological; it tends to reflect power. Yet, the creative ambiguity of language means that discourse can always be used to criticize the way power is distributed or even to propose an alternative system. Without the possibility of language, we are necessarily tied to reality as it is. Language may be a prison house, but it is one whose doors are left ajar.

My own work on technology and society has been profoundly influenced by the French historian and sociologist Jacques Ellul. He defines technique as the totality of methods rationally arrived at and having maximum efficiency (for a given stage of development) in every sphere of human endeavor.[2] I shall use the terms *technology* and *technique* interchangeably in the rest of the text.

The Technological System

Ellul realized that technique encompassed more than material techniques such as machines, engines, and factories. Nonmaterial techniques include techniques of organization, such as bureaucracy; psychological techniques, such as propaganda, advertising, and public relations; and the applied human sciences.

Two key ideas in the definition are those of efficiency and totality. Prior to the nineteenth century, technique was not dominated by the goal

2. Ellul, *Technological Society*, XXV.

of efficiency. The techniques of the past were as much an aesthetic expression as they were part of a search for greater effectiveness. Moreover, they were integrated into the larger culture and thus given ethical and religious significance. Technique did not dominate the culture.

In the nineteenth century, however, this begins to change dramatically. Technique in the above sense of efficiency becomes an end in itself. The West comes to believe in the myth of scientific and technological progress. This progress was supposed to lead to the happiness of the individual and the utopian perfection of the human race. Concurrently, it becomes more difficult to integrate the enormous proliferation of individual techniques into the culture. Technique slowly asserts its dominance over experience, tradition, and culture. Enter a technological civilization.[3]

With the proliferation of techniques and a conscious intention to create new and better techniques, technique became polyvalent. Machines, engines, bureaucracy (as a technique of organization), propaganda, the laser beam, and the computer proved to have countless applications. What eventually permits technique to become a system is the interrelating and co-ordinating of individual techniques. Machines can always be interrelated in linear fashion, as in a factory; but it is with the advent of the computer that each individual technique can become both an efficient, rational method and a source of information for other techniques so that techniques spatially separated can be coordinated. For example, in a large urban system, the computer permits the coordination of transportation, sanitary, medical, police, communication, and political subsystems. The real significance of the computer is not at the level of individual use; it is at the level of the organization of society. The computer has allowed technique to become a system and is, as it were, its nerve center.[4]

As both our life milieu and as a system, technique evolves at a certain price to culture, namely, the suppression of meaning. By meaning, I am speaking especially about meaning in a holistic and final sense: the meaning of life and the meaning of history, both of which can be expressed in either aesthetical or ethical terms and understood as either static or dynamic qualities. Technique attacks meaning in a number of ways. First, its widespread use in human interaction results in abstract and impersonal relationships. The other is the object of a technique that reduces everyone to the same abstraction. Technique becomes a substitute for manners and morality. Second, technique is solely responsible for the important actions that guarantee survival and provide a source of collective identity. The nontechnical actions

3. Ellul, *Technological Society*.
4. Ellul, *Technological System*.

we perform become either trivial or compensatory. Think of all the modern forms of escapist amusement—video games, drugs, sports, movies, tourism. Moreover, our own thoughts based on experience and belief are, next to technique, largely irrelevant as a source of motivation and as a limitation on our behavior; they too often become compensatory. For example, much of the recent work on ethics fails to take into account the technological milieu in which we are immersed and thus provides illusory solutions. As Ellul observes, in a technological civilization, everything becomes either an imitation of technology or a compensation for it.[5] Third, technique makes the meaning of keywords (aesthetic and ethical qualities) polyvalent. Although the words in all natural languages are to a certain extent ambiguous (the context reduces the ambiguity), polyvalent words are words so emptied of meaning that they can be applied to any situation or relationship. The psychological techniques of propaganda and advertising are largely responsible for this. Freedom and justice, for example, have been applied by politicians to the most tyrannical and unjust parties and governments. The decline in common meaning is accompanied by the impoverishment of language.

The technological system has been assisted in its growth by the proliferation and diffusion of visual images, especially those in the mass media. The materiality of visual images provides an excellent means of directly representing technique, which as a logical, efficient method is ultimately abstract; moreover, this same materiality provides a concrete context for our lives, something technique cannot do. It is no secret that we live in a civilization inundated by visual images: photographs, computer images, television, billboards, newspapers, magazines, comic books, advertisements, museums, movies, audiovisual aids, and so forth. Visual images are an excellent medium for arousing emotion; they can also be used to convey information. The communication of meaning, however, only occurs through language.[6]

The difference between the symbol and the visual image is fundamental. The symbol imbues reality—that which is empirical and quantifiable—with meaning. It allows us, for example, to make distinctions between good and evil, the sublime and the grotesque, freedom and tyranny. The abstract meaning(s) of the symbol allows one to draw a sharp distinction between sign and reality. In this view, the signifier or sign (spoken or written word), signified (meaning), and referent (external reality) all take part in discourse. Discourse, as language in use, is based on human experience and interaction with an external reality. If there were no referent, no common external reality, I would have no intersubjective experiences to be expressed in language.

5. Ellul, *Perspectives on Our Age*.
6. Ellul, *Humiliation of the Word*.

The visual image, by contrast, is a sign without meaning, a signifier without a signified. Visual images are typically about reality (a referent); moreover, the materiality of the visual images blurs the distinction between image and reality. Hence, visual images are conservative, for they tie us to reality as it is or is possible (always a technological possibility). As E. H. Gombrich argues in *Scientific American*, visual images only have meaning when the audience can supply the linguistic code and context for the image—the cultural context, the traditional verbal meanings the image has assumed.[7]

The visual images of the mass media are subverting language today. This is not mainly a consequence of the sheer quantity of visual images we encounter daily. Rather, it is due to a conjunction of three factors: (a) the logical linking together of the visual images in the media to form a pseudo-language; (b) the visual images of the media as a more emotionally satisfying reality than the reality we experience directly; and (c) the impoverishment of natural language.

The visual images of the media decontextualize reality, for they take reality out of its existential, cultural, and historical contexts. The visual images press upon us a strictly material reality, but one selectively represented. The visual images cut up and reconstruct reality according to a spatial logic of signifiers. The meaning of a television program is the final outcome—success or failure, survival or death, possession or nonpossession—in other words, power. If we look only at the visual images (turn off the sound and refuse to supply the missing context), we discover only images of power or images of possessions. We see objects to be possessed and consumed, the power of which becomes the power of the consumer—sports cars, cologne, computers, people; likewise, we see objects acting in relation to other objects—acts of possession, consumption, control, or violence. These acts of power are spatially linked together in a television program or movie in a logical sequence that leads to an outcome—success or failure.

The visual images of television are the most devastating because they are the most time-consuming and because their emotional effect is so great. Studies measuring the activity of the left part of the brain involved in rational thought indicate that when one is reading or listening to a lecture, that part is working so hard that one often feels tired afterward. By contrast, when one is watching television or a movie, the activity level in the left part of the brain is about what it is when one is asleep. It is obvious that the visual images of the media are working on us at the level of instinct and emotion. This is what makes the mass media such powerful compensatory

7. Gombrich, "Visual Image."

mechanisms. They provide pleasurable spectacles: life on television is emotionally more satisfying for many of us than the truth of our lives. Studies of soap operas indicate this, but the assertion applies to virtually all kinds of programs. The cumulative effect of technique and the universe of visual images are to turn us into spectators of life.

Baudrillard argues that reality is now simulated through the visual images of the media, computer models, and technical information in general. In this view, the visual image as sign has no meaning and no referent. It has swallowed up both meaning and reality. Baudrillard is a master of hyperbole. Although his assessment does not perfectly fit the present, it does ring true in several ways. First, many people tend to see the images of the media as only "signals and appearances" because reality (largely technical) has become more abstract, impersonal, and distant. Images, in this instance, are not of reality but the only reality.[8]

Second, reality itself is coming to reflect the visual images of the media. As Ellul notes, "Visual images are norms in a world without meaning."[9] Advertising, television programs, and movies provide models to be emulated. For young people, perhaps only slightly more than for adults, the media form the context of their lives. Rock stars, movie stars, and celebrities provide figures to be imitated. We remember advertising jingles, we dress like celebrities, and we speak and act like our television counterparts.

The impoverishment of language is both cause and effect of the triumph of visual images. Related problems include the reduction of language to a conduit of information and the reduction of language to a vehicle of emotional identification. The former problem is a direct expression of technique. The various expressions of this include the following: a growing scientific and technical vocabulary where the primary concern is functional relationships, operationalized terms, and quantified facts; the decline of metaphor in everyday speech; the equation of a word with a visual image or, in other words, the using of the visual image as the operational indicator of the concept behind the word; and the proliferation of morally and politically inoffensive words (retarded children are now exceptional children; the poor are underprivileged; immoral behavior becomes deviant behavior; old people are senior citizens). The latter problem—using words in vague ways for the purpose of emotional identification—is an indirect reflection of technique. Jargon, whose superficial appearance is scientific, is actually the opposite of rigorous, scientific, and technological concepts, for it is used less to convey information and more to sustain a feeling of

8. Baudrillard, *Simulations*.

9. Ellul, *Humiliation of the Word*, 147.

collective identity, a sense of belonging. Harold Garfinkel's study of mental health jargon in a psychiatric bureaucracy showed that in creating a closed linguistic community through jargon, mental health workers protected themselves against outside interference.[10]

Language that is used to convey information or to express emotional identification is language poorly suited to communicate aesthetic and ethical meaning. Technical and emotional discourse alone can never raise the question of truth (even if only understood as wisdom) over against reality.

The Technologization of the University

How have the technological system and the universe of visual images influenced the modern university? Let me begin to answer this with the following description of the modern university. Please compare it with your own. From a university that was on occasion and in some modest way a community of scholars—masters and students—we have come to an immense organization whose twin purposes are growth and efficiency. This bureaucratic organization that still goes by the name of university has a number of competing layers: at the bottom level, there is an earthly host of disgruntled degree and job seekers who regard reading, writing, and thinking as an obstacle to a life of security and comfort; at the middle level is a large stable of bureaucratic specialists whose chief goal is to get published and obtain research grants quite apart from the seriousness of the endeavor; at the upper level is a plethora of managers so concerned with public relations, dollars, and accountability measures that they have conveniently forgotten about the content of what is thought, learned, and studied, that is, the life of the mind. There are exceptions at all three levels, of course. But, as with any organization, most of its members are in conformity with its structure and believe its ideology.

The modern university has become an organization dominated by technique and the quest for a purely quantitative efficiency. When one examines the teaching-learning experience, the one issue that most stands out is the large lecture class with its attendant multiple-choice exams and decontextualized textbooks. Anyone who has taught such a class in an era where the onus of responsibility for learning falls principally on the teacher (an example of an unintentional dehumanization of students by stripping them of their responsibility) knows that the emphasis is on memorized facts. Anything complex or dependent on a demanding set of arguments is avoided. To teach hundreds of students with a minimum of

10. Garfinkel, *Studies in Ethnomethodology.*

misunderstanding, one must teach to the lowest common denominator. In large lecture sections, students are often not given any writing assignments—at best, a token assignment. The textbooks compound the felony by presenting decontextualized information for easy digestion. Textbooks have their place in mathematics and, to a certain extent, in the physical sciences, but in the humanities and social sciences, they are deadly. Students must place the "information" in these fields into a historical, literary, and philosophical context. And that means reading real books.

The large lecture section is not a recent innovation, but its combination with multiple-choice exams and textbooks written at a high school level or lower are of recent vintage. Textbook publishers are notorious for their concern that textbooks not challenge students, that they be "accessible." When a lack of writing is combined with a dearth of serious reading, the student's critical faculties are left dormant. Students discover that higher education means memorized information. This situation sometimes continues into upper division courses. I have found that many students in essay exams give me "grocery lists," that is, a series of facts presented in outline form. These students have learned to study for a multiple-choice test in which understanding is deemphasized, and now that they are expected to write an essay that should form a unity of understanding and reflection, they are befuddled.

This educational technique—large lecture sections with multiple-choice exams and textbooks written at a low level—is not just for the ease of teaching; it is likewise for the ease of learning. The communication of ideas, real education, is never efficient; it demands reflection, dialogue, and growth on the part of both students and teachers. By contrast, the dissemination and memorization of information can be made efficient. In 1970, two radical students asked me what could be done to shut down Illinois State University. My response was to take their education seriously. The system, I said, is premised on the assumption that the great majority of students are not really interested in obtaining a serious education. If all 17,000 students were to think about what they were learning and demand frequent dialogue with their teachers outside of class, the system would collapse.

As part of the increasing technologization of the university, there is the utopian quest for the perfect teaching technique. What interests me is that the quest is about motivation rather than course content. It involves a recognition that most students are here against their will, and in this sense, the university is but an extension of high school. The "in" technique today appears to be the extensive use of groups—group projects, group discussion, group papers, and so forth. The most relevant issue here is what effect this technique has on the content of the course: is there, as I suspect, a diminution of the difficulty of reading assignments with this method? Perhaps the advocates feel as most democratic peoples do, that there is more wisdom in a committee than in an individual. But all real learning is fundamentally an act of the individual.

I have nothing against the attempt to motivate students. Any teacher who ignores this altogether is arrogant. Still, it is of considerable interest that the emphasis on motivation in the quest for the perfect teaching technique simultaneously solves another dilemma—popularity. Student evaluations measure little other than how well liked a teacher is. Therefore, the motivation of students guarantees that they have momentarily become learners and that you the teacher will be well rewarded on evaluation day. The emphasis on motivational technique as an end in itself ignores several relevant matters. If the content of a course fails to motivate a student, one might plausibly assume that there is something wrong either with the student and his or her social environment—family, peer group, society— or with the course itself. With the correct teaching technique, however, course content and social environment can be ignored. Teaching becomes psychological technique.

Increased specialization is a necessary component of a technological civilization. Professors and their graduate students focus their attention on ever more refined research problems. The apparent gain in depth is offset by a loss of breadth. This carries over to the classroom so that course definition largely follows research specialization. Students learn more about their teachers' narrow interests and concepts than the reality to which they point.

It is in graduate school that future professors learn that research and publications are what matters. This is how one advances professionally. The great emphasis on research and research funding illustrates well one way in which the modern university is a part of the technological system. The technological system is an entirely dynamic system that evolves through innovation—more efficient techniques drive out less efficient ones. The funded research of the university, for the most part scientific in nature and technological in application, is a necessary prerequisite for the system to continue to grow. Even social science research is funded primarily to help solve social problems—a kind of social engineering. If in the past universities were subject to religious and political pressures, today these pressures are technological.

In the late 1960s and early 1970s, many serious objections were voiced about the moral and political implications of certain government-funded and privately funded projects—from Project Camelot to research in genetics. There is little opposition today. Research funding is an end in itself.

At what point does the proliferation of technical information become so overwhelming that it cannot be integrated? President Gregorian of Brown University, in an interview with Bill Moyers about fifteen years ago, spent considerable time on this problem. He observed that much of this information is not worth learning and that university teachers face an enormous task today in helping young people sift through this maze of information to make theoretical connections so that they are not left with a severely fragmented understanding of the cosmos, history, and culture.

There is no doubt that the effect of technique has caused universities to give up a common curriculum and proliferate the number of colleges and majors. The university accommodates itself to the technological division of labor in society. To be relevant it must do so. The modern university is a vast vocational training ground. I recognize that engineers, doctors, architects, and such must have an intensive technical education. My objection is to the technical takeover of the university's curriculum. But it is not the vocationalism in and of itself that is the key issue here. Remember that the medieval university was vocational. It is the meaning of vocational that is at stake. Vocational in the Middle Ages meant training in logic, oral expression, and later the composition of letters. There was a heavy emphasis on logic, grammar, and rhetoric. Today, vocational means technique. A technical education is the opposite of the theoretical education. A technical education means learning how to perform a task efficiently; it involves a transfer of information. A theoretical education involves understanding why, reflecting on explanations, and criticizing extant theories and the realities to which they refer.

In addition to the modern university losing a core curriculum (more about this later) and proliferating areas of study in vocational areas, the arts and sciences themselves have become exceedingly technical. First, they are highly specialized, one important aspect of technique. Second, they tend to emphasize the factual to the extent that concepts introduced rarely get beyond empirical concepts, that is, operational concepts. Third, courses proliferate that teach one to use this empirical, practical knowledge in a way that will attract prospective employers. We even have a major in public relations. All of this is at the expense of theory, which more and more becomes segregated into a course or two.

This brings us to another difference between a vocationalism steeped in language skills and a vocationalism determined by technique. It is only the former that permits one to raise the question of virtue. To raise the question of virtue without a strong foundation in language, history, literature, and philosophy is perhaps illusory. It is actually technical vocationalism that supplants whatever moral unity universities once had. Technique, as it were, deconstructs the university's moral meaning, which was, in one sense, embodied in its core curriculum. The core curriculum was what the university was about. When it worked well, the core curriculum was concerned with a pursuit of truth, in the form of wisdom, rather than with mere indoctrination into a static truth. Whatever the case, the university, until well into the nineteenth century, existed within a civilization whose institutions were structured according to symbolically mediated experiences.

A core curriculum is irrelevant in a technological civilization. The technological system makes symbolically interpreted experience and moral wisdom irrelevant. Yet, in another sense, we actually do have a core curriculum today, one that is concealed beneath the appearance of diversity and choice—computer literacy, quantitative methodology, and administration. Almost every major demands that its students be well versed in the use of the computer; many demand that students learn a logic of quantitative methodology; and increasing numbers want their students to learn the technique of administration—public administration, educational administration, hotel and motel administration, hospital administration, business administration, social work administration. The emphasis on the computer, quantitative method, and administration means that the curriculum becomes increasingly similar at a deeper level than that of its apparent diversity. There is little difference between business and educational administration insofar as these are empirical and not normative disciplines. The reason is obvious: technique in the form of bureaucracy or as an expert system renders all organizations the same.

This brings us to the administration of the modern university. If teaching and learning have been brought under the aegis of technique, administration has been even moreso. Bureaucracy is a technique of organization. Its primary concern is efficiency, and as such it is amoral. Its rules are rational, technical rules, not moral rules. Bureaucracy thrives on the quantification of tasks. For action to be reduced to a technical task, it has to be rationalized and quantified. This quantification of behavior becomes the backbone of accountability measures. Hannah Arendt realized that the structure of bureaucracy at one level gives rise to the bureaucratic mind at another level. The bureaucratic mind perceives the world as an object to be manipulated, on one hand, and as a set of technical rules to be conformed to, on the other hand. Arendt saw in the case of Adolph Eichmann an extreme example of what a bureaucratic mind can do to one's moral sensibility.[11]

Bureaucratic control is dependent on accountability measures. The most telling measure is the student evaluation, which, as most of us know,

---

11. Arendt, *Eichmann in Jerusalem.*

measures popularity. But for the bureaucrat, who has neither time nor the inclination to define good teaching, this provides a quantitative basis for judging teaching. The use of student evaluations has no doubt played a large part in grade inflation. Teachers must now placate students by making learning fun (if there is one thing real learning is not, it is fun) and by giving higher grades.

Perhaps this is only retribution for a grading system that has become a technique of control. Most grading systems in this country range from a scale of five to nine letter grades. The assumption, I presume, is that each grade represents a true qualitative difference. I am not convinced that we can measure intelligence or performance in any real sense of the term. But my point is how grades become for the teacher a bureaucratic form of control, a substitute for authority and, for the student, an end in itself. Along with grades, consider the proliferation of tests and busy time assignments, all of which are to be graded. These have become forms of coercion. We are afraid that students will not take us seriously if we do not give a large number of tests and assignments. Students actually desire these coercions, for they are addicted to grades, the way we are addicted to student evaluations, merit ratings, and the like. A perfect indication of the bureaucratic mind is that we have accepted the accountability measures that routinely strip us of our freedom. Some research indicates that administrators use measures of performance, even when their validity is seriously questioned, because of their accessibility.[12]

I sometimes reflect on my experiences in Ireland in 1968. At that time, Ireland had not converted to the American system of individual courses and tests for each course. In that system, modeled to some extent on the British system, students only took exams at the end of the year and these only in their major and minor areas. When I explained how the American system worked with numerous tests and assignments in each course, their reaction was that this was a form of punishment, on one hand, and an obstacle to learning, on the other hand. It's not intended as punishment, but it is certainly an obstacle to real learning. Some of my best students, upon completion of their senior year, comment, "At last I can read something and attempt to understand it."

The quality of the teaching-learning experience is indicated less by the quantity of tests and assignments, by the quantity of the reading assignments, and by how hard a grader the teacher is than by the content of the course and the character and intelligence of both teacher and student. When higher education was not primarily technical, and thus more theoretical,

12. O'Reilly, "Variations in Decision Makers' Use."

the ideal teacher was one who had both breadth and depth of learning and the ideal student one who desired to attain both character and intelligence. Those teachers who still emulate this ideal are ignored by the majority of students who simply desire a technical education—hence, the escalation of tests and assignments, grade inflation, and the use of student evaluations. Education has become a ritual in which each moiety exchanges gifts—high grades and entertainment for high evaluations. Those teachers who do provide students with courses in their majors designed to help them obtain a job meet with a much better response. But there is a problem that transcends both theoretical and technical education—the declining ability or at least motivation of students to read and think. (I'll come back to this in a moment.) Scholarship has likewise been adversely affected by bureaucratic technique; it has become research productivity. The name of the game is publications and research grants. As repeatedly reported in *The Chronicle of Higher Education*, there has been a steady increase through the years in the number of publications expected of faculty. Productivity as an end in itself, quite apart from a sense of the seriousness of the research endeavor, is not a matter of professional responsibility but of bureaucratic necessity. Upwardly mobile universities committed to growth demand more research because this brings the university more dollars and prestige.

Taken together, the various accountability measures make up the basis for the yearly evaluation process, which, once again, is not a matter of professional responsibility but of bureaucratic control. I am not arguing against making rigorous tenure and promotion decisions but only against the irrationality of yearly merit for intellectual labor. Merit should be exclusively tied to promotion for faculty to work on large and serious scholarly activities both in and out of the classroom. Perhaps the bureaucratic mind is nowhere more in evidence among faculty than when they serve on committees at the college and university levels. Their attitude often becomes one of organizational seriousness—they become administrators. For the most part, these committees make suggestions about changes in bureaucratic procedures that are intended to make the university more efficient. For example, on curriculum committees, whether the content of a proposed course is serious typically gets less attention than whether the course write-up follows bureaucratic regulations. Managers and their committees always think about structure, rules, and procedures. But all nontechnical change occurs at the more humble level, that of people. A university that truly wishes to enrich the teaching-learning relationship and to enhance serious scholarship accomplishes this by giving faculty and students the freedom to be serious. In a word, it pretty much leaves them alone. A university that attempts to change itself through bureaucratic reform and public relations only makes things worse. Today, we are being overwhelmed by bureaucratic procedures and accountability measures, like the truly silly idea of value-added education. We spend so much time accounting for what we do that we don't have time to do what we do. From the point of view of traditional university, every attempt to reform the modern university has only made it worse—that is, brought it more into conformity with the technological system.

What role has the universe of visual images and the impoverishment of language played in the transformation of the university? Nothing is more impoverished than bureaucratic language, which is, according to Ralph Hummel, a dead language.[13] There is no real communication in bureaucratic language, only a transfer of information. Natural languages that are still living languages permit the speaker to communicate intentionality (not perfectly, of course) and to make reference to a world that is both empirical and normative. Bureaucratic language does not express intention and at times does not refer to a world outside of itself. It is hermetic in both major senses of the term.

On its manifest side, bureaucratic language is a language of functionality not in the sense of purpose but of consequence. It is a language whose meaning is made up of fragments of operationalized information that are used to solve problems. Bureaucratic language may be used to describe organizational procedures or be used to give commands to follow such procedures. Because a bureaucracy's very reason for existence is efficiency, its language is about efficiency—functions, operationalized indicators, technical solutions. Computer language is merely a variation of bureaucratic language.

The tacit side of bureaucratic language is its vagueness with respect to aesthetic and ethical meaning. This language is a language designed to represent things, logical procedures, and functions. It is not a language that permits one to refer to qualities that are essentially metaphorical. This is why university committees and administrators discuss the values, goals, and mission of the university and make "vision statements" in such a vague way. The language they employ is not suited for the discussion of value, mission, and goal. Bureaucratic language is precise about technological function but vague with respect to meaning. The vagueness of bureaucratic language is an example of jargon I discussed earlier. Bureaucratic discussions of values, goals, missions, and so forth allow the university to achieve a purely emotional identification on the part of faculty, students, and administrators. For who could be opposed to a liberal education, humane studies, becoming a "premier undergraduate institution," or "building pre-eminent graduate

13. Hummel, *Bureaucratic Experience.*

programs"? What these words mean anymore is anyone's guess. Therefore, the very vagueness of keywords permits the university to do whatever it pleases, even if it means further distancing itself from a liberal arts ideal, and yet remain under the emotionally integrating aegis of an impoverished vocabulary. As this kind of language penetrates everyday life to a greater degree, students are less able to understand more rigorous, theoretical language. As I indicated earlier, the universe of visual images in the mass media has had a profound effect on the motivation and ability to think abstractly, to read, and to write. Those who have been the heaviest television viewers have the greatest difficulty. The question of motivation is easier to deal with than that of ability. Students today would rather surf the net, watch television, go to movies, and play video games than read and think because the former activities are more immediately pleasurable and emotionally satisfying. Our culture, moreover, defines the purpose of life as consumption and spectacle. The context of students' lives is the media, not the world of books and serious discourse. For a student not to be in touch with the mass media is tantamount to being out of touch with reality.

The declining ability of students to think abstractly is related to the fact that to understand phenomena visually requires a different process and operates at a different level. Visual images hit us at an emotional level (as was indicated earlier) and are understood through association with other visual images, with the result of an overall, intuitive apprehension of empirical reality. By contrast, words permit us to think abstractly so that words are related to one another to form larger units of meaning—sentences, paragraphs, narratives. But the meaning is abstract. With visual thinking, the meaning is concrete.

Now humans have always had visual images and these have always formed a background to our mental images. Today, however, we are engulfed by visual images—television, movies, billboards, photographs, pictures in magazines and newspapers, posters, and video games. The sheer volume and speed with which they bombard us is less important than their being logically related in a television program, a movie, a slide presentation. Even when words are used in conjunction with the visual images, they are subordinate to the images. The word is losing its meaning and value as a vehicle for reasoning; instead, it has become an "accessory" to the image, the operational indicator of the word. One of my most insightful students has written about how music videos have invaded his memories of adolescent romances. Popular songs, the lyrics of which symbolized at the time his romantic relationship, now evoke memories of the accompanying video even more than they release memories of his actual girlfriend.

This is why, I am convinced, many students remember the example and not the concept, only remember what is written on the blackboard, and can't follow a logical argument. They demand and we provide audio-visual aids, movies, slides, charts, visual models, and even more pictures in textbooks. What we don't realize in our sincere effort to motivate students is that we are concurrently weakening their ability to think abstractly by subjugating the word to the image. Moreover, as visual images proliferate as instructional aids and as bureaucratic language increasingly becomes the language of instruction in classroom and in textbook, the university is further deconstructed. It is being robbed of its possibility for meaning, that is, the possibility to raise the questions of virtue, wisdom, and beauty with students in a critical way.

There are some indications that the decline in the ability to think abstractly is greater with respect to verbal skills than with respect to mathematical skills. Mathematical symbols are purely imaginary; they have no human context. Verbal symbols can only be understood in context. For students to understand the meaning of keywords and to be able to think abstractly through language, they need to know the historical, literary, and philosophical contexts in which these concepts and metaphors are used. As I have already indicated, cultural literacy is of little practical consequence in a technological civilization. It was of value in civilizations whose institutions were symbolically mediated. In a technological civilization, the literacy demanded is computer literacy.

The computer poses as great a threat to language and the ability to think dialectically, as do the mass media. At the level of individual use, the language of the computer is exactly what I have described as the manifest side of bureaucratic language: a rationalized language of information. The great danger to education is not so much that students learn to use the computer (for they must) but that more and more communication between humans will be reduced to information transfer from human to computer and vice versa. The arts of dialogue and writing will be seriously impaired.

The computer poses as great a danger to thinking as it does to language. The logic of the computer is a logic of mutually exclusive categories. But the logic of existence is dialectical. That is, life as it is experienced individually and collectively is rife with ambiguity, ambivalence, and contradiction. Poetry best expresses this dialectical perspective, but all the humanities and social sciences require a dialectical logic to remain authentic. It is only the dialectical perspective that allows one to develop a critical mind. The computer allows us to impose a monolithic logic and quantitative methodology on all of life, to reduce the qualitative to the quantitative, to reduce communication to information, all in the name of efficiency. The computer is a threat to individual and collective memory. Our memories decline to the extent that everything is done for us (the computer) and shown to us (the images of the media). Memory centers on events that have acquired emotional, intellectual, or moral significance to us. The past understood in either mythological or historical terms acquired meaning through key events that took on symbolic meaning. The information from the computer, as we have previously seen, is abstract, logical, and quantitative but ultimately meaningless. The cynical worldview of the computer is that we live in a random and meaningless world about which the omnipotent computer can generate an infinite amount of information that we can exploit to our advantage. The use of the computer in direct instruction should be sparing indeed.

The Need For A "Shadow" University

The liberal arts survive as a distant and vague memory that still haunts us. When teachers don't teach the liberal arts in broad fashion, when students don't want to learn them, when scholars become too specialized and technical in their approach, and when administrators cease viewing administration as a necessary evil and the least important thing a university does, then how can they thrive?

If I am correct in my assessment, we can't have it both ways: either a university with a strong core curriculum devoted to history, literature, philosophy, and the social sciences that has as its aims both an appreciative and critical understanding of our civilization, facility in the use of language, and the search for civic virtue; or a university oriented to turning out technicians. Only the latter type of university is truly efficient; it becomes a factory that mass produces its student products—they all become technicians, managers, and most of all, short-term memorizers.

It all boils down to this: we have chosen the latter, the university without qualities. Robert Musil's great novel, *The Man Without Qualities*, is instructive here. It depicts in the character of Ulrich, the plight of modern man reduced to the status of a mere functionary, one who acts but whose subjectivity is in the state of flux; one whose self is without a moral core. The man without qualities does not stand for anything. The modern university does not stand for anything either—except efficiency and growth, or in other words, power. The university has become an institutional functionary in the technological system. If at times the university has been the most open of our institutions, today it is not. If we still believe in academic freedom, then one of the most important acts of freedom will be to oppose the monolithic technological structure and content of the university.

Critical analysis demands practical suggestions. What can be done? First, we must start with a realization that the modern university is unreformable. The university I have described is by and large in harmony with the development of the technological civilization. The university as institution will not change radically until the larger society and civilization do. To reform an individual university in the direction I have suggested would be sheer folly. Legislators, parents, and the students themselves would not stand for it. You can have a Great Books of Western Civilization-based college here and there as long as it is the rare exception. Please don't think, by the way, that I am recommending a Great Books approach. To me, a Great Books curriculum as a total project ossifies the past.

Second, we can all act as individuals against the technologization of the university. We can refuse to be mere specialists by reading broadly, thinking deeply, and searching for the connections so that we develop a better sense of history and culture. We must all become historians, literary critics, philosophers, and sociologists. This continuing self-education should be incorporated into our teaching and our scholarship. Moreover, we can refuse to be lackeys of the bureaucratic structure by refusing to serve on most of the college and university committees.

Third, we must refuse to teach to the lowest common denominator, for we must make the motivated student our chief concern. We can't truly educate all the degree seekers. Public opinion polls indicate that the overwhelming majority of college students never read a serious book after graduation; some managed to avoid this during college. If college is just a start on a lifelong education, then we are dismal failures. Memorized information is retained for a short period of time. For multiple-choice exams, textbooks, and large classes, this is the medium of exchange. Why kid ourselves? The majority of students are going to retain little of traditional value from their college education and will not continue to educate themselves. They will leave here with credentials and a few technical skills.

And yet, we can't abandon the unmotivated majority altogether. Their parents have paid for their education. Some of these students will even be turned on to serious learning at some point in their college career. The best we can do within the current university is fight a holding action. Find a way to pitch the course at a level that challenges the motivated students and yet allows the normal run of students to pass. This can be done in part by adjusting the difficulty of the test and one's expectations to this average student.

Finally, if the current university is unreformable and the actions we take as individuals are less than fully satisfactory, there is another way. We can, in concert with other like-minded individuals, offer courses to college students and adults in the community outside the usual structure. I envisage that these courses be offered without payment and without grades. There still could be evaluation, but this could be done without assigning a letter grade, simply through the remarks one makes. These courses could be individually or team taught. The point is that, for the most part, they should be broadly defined, interdisciplinary in outlook, and either appreciative or critical in tone. There is no reason that some not be designed to appreciate the cultural heritage, whereas others be intended to offer a critique of this heritage especially as it applies to modern society.

Because no grades, no degrees, and no money would be awarded, this shadow university of ideas would be completely gratuitous. No one would be either a student or a teacher unless one desired to teach and learn in this way. The classes would be small and more discussion-oriented than the usual class.

The freedom of this enterprise—the freedom to teach without bureaucratic regulation, to teach what one wishes and how one wishes without regard to popularity and salary, and the freedom to take a course without concern for money, degree, and job—would serve as a vital criticism of the modern university, whose main function is to help us conform to a technological civilization, at the price of language, criticism, and virtue.

Chapter 7 —————————————————————

Ethical Individualism and
Moral Collectivism in America

Originally published in *Humanitas* 16, no. 1 (2003) 56–73.

From Alexis de Tocqueville to Robert Bellah and Alan Wolfe, the many observers of the United States invariably call attention to its emphasis on individualism. In the popular culture of American television and movies, the autonomous individual stands out, whether as rebel against the system or as self-centered consumer of endless products, services, and other people. Then there are the advertisements that tell us how important we are as individuals: "you are worth it." Over against these seemingly positive images are those of individuals worried about their personal relationships—lonely, depressed, and forlorn. But everywhere we find individuals thinking about and acting for themselves.

The way ethics is taught in the public schools in the United States does little to dispel this idea that individualism is the hallmark of American culture. James Davison Hunter's recent study of moral education confirms this.[1] He identifies three approaches to moral education in American public schools: neo-classical, communitarian, and psychological.

The neo-classical position is similar to that of natural law theories in earlier centuries. It maintains the existence of universal moral values or virtues, whether that universality as ascertained by reason derives from nature or history. The communitarian approach emphasizes the consensus in the community about what is moral and what is immoral. This consensus should come from democratic participation rather than be imposed from above by an elite. The needs of the community should take precedence over those of the individual.

---

1. Hunter, *Death of Character.*

The psychological approach to moral education is dominant. It is loosely tied to the "self-esteem" movement, which stresses the therapeutic function of education. The purpose of education is as much to make the student feel good about himself as to educate him. In one version of the psychological approach, moral judgments become "value preferences" and values hardly more than individual emotions. The hope is that through self-expression and interaction with others the student will clarify what his values are. In a sense the values are already inside the student and simply need to be teased out. Another version of the approach emphasizes reason and turns morality into a means for individual success or happiness.[2]

In perhaps his most important finding, Hunter observes that despite their manifest differences all three approaches share the same "assumptions, concepts, and ideals."[3] The reason is that each of the approaches takes morality out of its cultural context and thus renders it abstract. Moreover, morality is presented to the students as subjective. With the triumph of the scientific worldview, objectivity became identified with scientific inquiry; religion and morality in turn became subjective.[4] As Louis Dumont[5] observes, the modern ideology turns morality and virtue into personal values that individuals are free to accept or reject. Emotivism as a moral philosophy is the academic recognition of morality having become a consumer choice.[6]

In the long run, however, moral education programs are not successful, not even those that occur in religious schools. They are not able to counter what Christina Sommers calls the basic assumptions of students entering college: psychological egoism (motivation is invariably selfish), moral relativism, radical tolerance, and moral responsibility centered in organizations not individuals. This final assumption is telling. As Sommers demonstrates, the main thrust of moral education beyond self expression is to become interested in the social policy of private and public organizations so that one chooses the right organization to solve the problem.[7] Students become at worst "moral spectators" and at best activists, but without a sense of personal moral responsibility. Morality is either emotional preference or political preference or both. The widespread idea that moral responsibility resides in society and society only is a sign that the individualistic approach to morality in the United States may not be what it first appears to be.

2. Hunter, *Death of Character*.
3. Hunter, *Death of Character*, 122–27.
4. van den Berg, *Changing Nature of Man*.
5. Dumont, "On Value."
6. MacIntyre, *After Virtue*.
7. Sommers, "Ethics Without Virtue."

Psychological Weakness

The theory of the mass society is one of the great theories in the social sciences, I think, and unfortunately one that has been more or less abandoned. It was central to the work of C. Wright Mills, but much earlier, in the nineteenth century, it appeared in nascent form in the work of Kierkegaard[8] and Tocqueville.[9] A mass society is one that possesses simultaneously a high degree of collectivism and individualism. The medieval state and church exercised little direct control over local communities except in times of crisis, e.g., war and heresy. Even if they had wanted to, transportation and communication would have made it difficult. With the growth of power in the state, administration, and technology and the great increase in migration and social mobility in the eighteenth century, the local community began to decline as an agency of moral control, a development accompanied by the gradual dissolution of the institution of the family. The abstract power of the state, administration, and technology supplanted the moral authority of the community and family. In Dumont's terms, "hierarchy" had previously been set within "holism"; that is, the interests of the community took precedence over the interests of those who had higher status and power in the social hierarchy. With the rise of the values of equality and individualism in modern ideology,[10] hierarchy becomes unstable and difficult to justify except in terms of individual competition. Therefore centralized power that affords everyone equal treatment appears an attractive alternative to hierarchy.

But there is a price to pay for equality. That both citizen and government favor equality, Tocqueville noted, should alert us to the danger.[11] The state administratively applies the same regulations to everyone. The state, administration, and technology are considered efficient to the extent they can impose the same abstract rules on everyone.

The individual citizen reacts ambivalently to the decline of personal and cultural authority, Tocqueville tells us. It produces feelings of release, freedom, and power. No one can tell me what to do, for we are equal. At the same time, however, we cannot rely on others for assistance; they are not morally bound to us in a reciprocal relationship. Moreover, our relationships to others become more competitive, more dangerous. This leads to what Tocqueville terms "psychological weakness." We live in tacit fear of others, not so much of their potential for physical violence as of

---

8. Kierkegaard, *Present Age*.

9. Tocqueville, *Democracy in America*.

10. Dumont, "Anthropological Community and Ideology."

11. Tocqueville, *Democracy in America*, 672.

their ability to manipulate us. Trust presupposes a moral community, which in turn requires moral authority. Our relationships become vague because they are based on distrust. Individualism in this context involves psychological weakness; we look to the peer group and the state to protect us from exploitation. Psychological weakness, moreover, is the bridge between individualism and egoism (Tocqueville noted that the former eventually led to the latter.)[12]

Fragmentation and Depersonalization

Tzvetan Todorov's masterful *Facing the Extreme: Moral Life in Concentration Camps* discusses the fragmentation and depersonalization of the individual in the context of totalitarianism, of which concentration camps are its fullest expression. The state assumes control of all social goals and appropriates the individual's social existence. In effect, the individual is denied moral responsibility for her actions. Fragmentation and depersonalization make it difficult, if not impossible, for the individual to exercise moral judgment. Fragmentation and depersonalization, then, represent the internalization of totalitarianism. We will see later, however, that these twin psychological maladies occur today in less extreme contexts.[13]

Fragmentation involves the splitting of the self in a variety of ways, including that between the public and private spheres of life and between thought and action. Todorov provides the example of a Nazi guard who treats inmates in cruel ways at work but hours later is a kind and loving father to his children in the privacy of his home. Another is the inmate who retains his religious beliefs but informs on fellow inmates.

Manifestations of fragmentation in the modern world include technical and bureaucratic specialization and professionalization. Personal responsibility is narrowly limited to one's specialized function. No one person is responsible for a decision in the modern bureaucracy. Our responsibility is further diminished by our dependence upon specialized experts who have invaded all spheres of life.[14]

The more technology objectifies human ability and intelligence, the less one needs to rely on personal experience and tradition. It is easy to forget that technology affects us as we create and use it. What it requires of us in an age of instantaneous communication and action is reflex not reflection, as Jacques

---

12. Tocqueville, *Democracy in America*, 507.

13. Todorov, *Facing the Extreme*.

14. Todorov, *Facing the Extreme*.

Ellul has observed.[15] Our own thoughts are increasingly irrelevant and, in compensation, turn toward fantasy and illusion.

Depersonalization refers to the treatment of a human being as a nonperson. To handle a person as simply an inmate, or as an abstract category, to define someone exclusively in terms of statistical information, or to act as if she were less than human is to depersonalize the other. But depersonalization runs in both directions. Under totalitarianism everyone is turned into a "cog in the machine." This in turn results in the obedience to authority syndrome or the bureaucratic mind. When faced with unlimited or arbitrary power, one must submit to it. Totalitarianism deprives individuals of their will. As Todorov observes, "each and everyone is both guard and inmate at the same time."[16] Once again Todorov compares the dehumanizing aspects of modern bureaucracy and technology to totalitarian practices. Technology and bureaucracy mediate human relationships, permitting a vast increase in extensive, abstract relationships in the interest of efficiency at the expense of intensive, immediate relationships.[17] Abstract rules govern virtually all human relationships in modern organizations, including the university. Teaching, for example, is moving in the direction of a technical and contractual relationship between teacher and student in lieu of an informal, human relationship.

If the bureaucratic mind leads to submission to authority, it also results in its opposite—manipulation of others. In a universe of raw power, one submits to a power greater than one's own and manipulates that of lesser strength. In such an environment moral judgment becomes progressively superfluous, for everyone perceives that others invariably act out of self-interest.

Not everyone views fragmentation and depersonalization as destructive, however. Some postmodernists celebrate fragmentation and depersonalization as expressions of individual freedom rather than as a consequence of extreme collectivism. Fragmentation permits one to escape the moral unity of the self and become a mere role-player, one who approaches life in an exclusively aesthetical and apparently free manner. Life becomes an experiment or a game, whose rules are controlled by the centralized power of the state, the corporation, the media, and technology.

Depersonalization is also sometimes perceived as a form of freedom. The spoken word, according to Jacques Ellul, is the most appropriate medium for sustaining deep human relationships and for making moral

15. Ellul, *Technological Society*.
16. Todorov, *Facing the Extreme*, 166.
17. Todorov, *Facing the Extreme*.

judgments.[18] Do you stand behind your words? Do you keep your promises? Advertising, however, is anonymous discourse directed toward an abstract audience of consumers. With the computer, everyone can actively engage in anonymous discourse and can say anything, no matter how preposterous and hurtful, without any risks. The computer encourages the most irresponsible discourse yet known. In effect, it teaches the tacit lesson that freedom exists without responsibility. Psychological weakness warrants such depersonalized discourse. Modern individualism entails a fragmented, depersonalized self living in fear of others. This is hardly the cultural ideal of the individual in the Renaissance or in the Enlightenment. Rather, when expressed in postmodernism, it is the ideological justification of an increasingly collectivized existence.

The Mass Media Aestheticize and Thus Fragment Life

In the subsequent discussion of technology in general and the mass media in particular, I am making several assumptions. First, technology is the fundamental basis of and paramount determining factor of modern societies. This determinism, however, is not metaphysical but sociological; as such it can be resisted and even overcome but only with great effort and keen insight. Second, the traditional relationship between language and visual images is gradually being reversed. In the past, the symbolism of visual art took its meaning from the semantic foundation of culture.[19] Discourse provided the context within which visual images assumed meaning. Today the opposite is becoming true. For increasing numbers of people, the images of the media furnish the context within which words take their meaning. Hence the visual images of the media, which are increasingly first related to one another before they are related to language, serve as "operational indications" of words. The reification of language results, for example, in the meaning of love being reduced to the image of an embrace or a kiss. The reification of language is furthered by propaganda, advertising, and public relations, whose use of words destroys their common meaning and thus renders them vague and ultimately meaningless.

Those most affected by the images of the media, research indicates, are those whose reading skills are poor, those who watch a lot of television (and related visual media), and those who are lonely.[20] Jane Healy has documented

18. Ellul, *Humiliation of the Word.*

19. Gombrich, "Visual Image."

20. Stivers, *Culture of Cynicism,* 157.

the serious decline in reading comprehension in the country.[21] There is thus little indication that the process of the subordination of language to images in the media will easily or soon be reversed.[22] In a sense, I am describing a future that is rapidly moving toward us.

The mass media reinforce and deepen the fragmentation and depersonalization that bureaucracy and technology unintentionally create. The aesthetical and the ethical, according to Kierkegaard, are dimensions of culture and ways of existing. Both are necessary to the life of the individual and society. The aesthetical is concerned with immediate experience. Aesthetic existence is principally about enjoyment, to lose oneself in the pleasure of the moment. A purely aesthetical approach to life, Kierkegaard observes, is ethically indifferent to others. When one is not ethically bound to others, one is free to relate to them as best fits one's needs and desires.

The ethical is concerned with responsibility toward and limits in our relationship to others. An ethical approach to life provides a moral unity to the self; one is the same person, no matter what the circumstances. When one stands for specific beliefs and puts them into practice one becomes a coherent, consistent, and unified self. To paraphrase Kierkegaard, one chooses a self for a lifetime. An aesthetical approach to life cannot provide unity for the self because there is no unity in pleasure. Instead pleasure entails self abandonment; one merges with the object of pleasure.[23] Therefore an exclusively aesthetical approach to life requires a multiplicity of selves—a different self in each situation.

Television and related media are, in their overall impact, antinarrative. (In this section I will use television to refer in a general way to all media that feature visual images.) Although it can be argued that individual television programs have a narrative form (even here I argue that in the electronic media the visual images destroy the narrative structure of discourse), the entire spectrum of programs is random and incoherent. That is, there is no temporal and meaningful relationship among programs and commercials. One can go from the news of an earthquake to a commercial for hemorrhoids, to a talk show about men who are looking for a mother figure in the women they date, to a game show, to a program that recreates "true" police encounters with criminals. Therefore television in its total impact destroys the experience of event time. One is left with duration time, the continuous time of description. Television describes reality for us but

21. Healy, *Endangered Minds*, 26–36.

22. See Stivers, *Technology as Magic* for a fuller discussion of these issues.

23. Kierkegaard, *Concluding Unscientific Postscript*.

leaves us with no understanding of it. The more television one watches, the more life appears absurd, but interesting.

The main, if not exclusive, impact of the visual image is emotional.[24] Emotional experiences are principally aesthetical, and as such leave us oriented to the moment of pleasure or pain. By itself emotion does not allow us to transcend the immediate present. What is most distinctive about humans, Kierkegaard argues, is our imagination and anticipation of the future; without this, there is no sense of the past.[25] Television's visual images permit no future and thus no past. Television creates an eternal present. To live exclusively in the moment, to live from moment to moment, is to live a fragmented existence.

Television makes a fundamental appeal to our instincts. In short, television's images are pleasurable. Paul Goldberger maintains that "[t]he rise in visual literacy has been accompanied by an almost desperate desire to be stimulated."[26] Our increased visual sophistication lowers our threshold for boredom; we require ever more spectacular experiences.

Television plays a large role in the representation of life as spectacle. According to Guy Debord, we now live in a world of visual representation, a mirror-world in which the image is more important than and indeed defines reality. Moreover, life has largely been transformed into an image for immediate consumption. The spectacle is the "language" of the commodity; it is the visualization of the commodity for spiritual consumption.[27] The spectacle serves to reinforce the extreme individualism of consumerism. I become what I see and what I consume. I possess as many selves as the products I consume. The media fragment time and the sense of a consistent, coherent self.

The Mass Media Objectify and Thus Depersonalize Existence

Television makes discourse anonymous. It is information sent by no one to anyone. It is impossible to trace the information back to a personal source, for even newscasters often work from scripts written by others. The audience is only a statistical audience of people with similar characteristics as determined by marketing techniques. Communication achieves its highest state

24. Gombrich, "Visual Image."
25. Kierkegaard, *Concluding Unscientific Postscript.*
26. Goldberger, "Risks of Razzle-Dazzle."
27. Debord, *Society of the Spectacle.*

of impersonality in the media. The depersonalized information of the media seems more objective than that provided by a person.

Walter Benjamin has called attention to the destruction of meaning that occurs when a work of art is removed from its historical and cultural context and is technologically reproduced exclusively as a visual image for consumption.[28] This objectification is essentially what television does on a much larger scale.

Television appears to be describing reality, particularly in news programs, documentaries, talk shows, and game shows. In effect, it is reconstructing reality by taking reality out of its temporal and cultural context. Reality as we live it still retains some meaning, no matter how small; but television expunges this meaning and recomposes reality as a sequence of image fragments. Television is anti-surrealistic, as Ellul notes; it subtracts meaning from life.[29]

A former graduate student of mine talks about a special song that his girlfriend and he shared. When he heard the song, he thought about her and their experiences together. Once he viewed the music video of that song, his images were altered. Now when he heard the song, the images of the music video appeared in his mind. His girlfriend and their experiences had vanished.

Poems, novels, and stories, by contrast, provide shared symbolic experiences to listeners and readers, which have to be filtered through the reader's own meaningful experiences. The media objectify our experiences and thus control them. Is this not a form of totalitarianism?

Modern individualism exists to permit a non-political totalitarianism to flourish. Consumerism creates a radical individualism. But as Baudrillard observes, our freedom as consumers to choose among a variety of commodities is set within the overwhelming constraint to be consumers.[30] Consumerism is a forced and total consumerism, a total mobilization of consumers. Psychologically isolated individuals are a precondition for non-political totalitarianism. Psychologically weak, fragmented, and depersonalized individuals are easy prey for the centralized power of the state, corporation, bureaucracy, and technology. If it were not for the mass media, modern individualism might appear impotent. The aestheticizing power of the media can magically transform weakness into strength. The mass media give us totalitarianism with a human face, a kinder and friendlier totalitarianism.

28. Benjamin, "Work of Art."
29. Ellul, *Humiliation of the Word.*
30. Baudrillard, "Consumer Society."

Extreme Individualism as a Necessary Component of Moral Collectivism

As we have already seen, human relationships today tend to be based on considerations of power or on aesthetical considerations of style and consumption. Modern societies make it both unnecessary and difficult for individuals to assume moral responsibility and exercise moral judgment. This is especially true if one considers the emergence of an ersatz morality that is thoroughly collectivistic; it assumes no moral judgment and requires no moral responsibility. Yet its effectiveness depends upon its apparent individualism. I have described this pseudo-morality, at least in its American context, in *The Culture of Cynicism*.[31]

Technology is one of the forms that modern morality assumes; its greatest influence is exerted through organizational and psychological techniques. Max Weber understood bureaucracy to be a kind of machine, a kind of technology.[32] Bureaucratic rules carry the weight of morality for those who possess the bureaucratic mind. Psychological techniques, at least those on an interpersonal level, are imitation technologies that promise an effective way to control the other. The countless "how to" manuals, guides, and books on everything from marital happiness and child-rearing to dressing for success invariably involve a set of steps or a logical process that more or less guarantees success. These become a substitute for conventional manners and morality. They turn human relationships into relationships of power. When we apply a technique designed for maximum efficiency to another it appears to be an expression of individual freedom; when it is applied to us, it registers as collectivistic, as beyond our control. Technical and bureaucratic rules require no moral judgment, for procedural rules do not depend upon context for their meaning as do traditional virtues and moral principles. Nor do they require moral responsibility, for responsibility is embedded in the organization and technology.

Public opinion is a kind of statistical morality in which the majority viewpoint or statistically average behavior becomes the norm. The normal begins to replace the moral with the onset of the belief in technological progress or at least the belief of radical immanentism, that is, that we live in a self-contained world without ultimate purpose. The normal, then, is either the moral in the theory of progress or a practical guide in a nihilistic world. The result is that public opinion becomes a statistical morality in which the normal assumes the guise of a moral norm. American historian Daniel

31. Stivers, *Culture of Cynicism*.
32. Weber, *Economy and Society*, vol. 2.

Boorstin discusses how middle-class parents in the early twentieth century came to be greatly concerned with their children's measured development in intelligence, personality, and behavior. Parents wanted their children to be normal, to fit in.[33] As Tocqueville noted, public opinion becomes tyrannical in a democratic society because citizens fear being isolated if they dissent from it.[34] This is a variation of Tocqueville's earlier psychological weakness argument. Here it is not the fear of being manipulated, but rather the fear of isolation. Parents feared for the isolation of their children. At the same time, however, we are told that the expression of our opinion in polls is a form of freedom. Public opinion polls flatter us. Public opinion likewise acts to demand that those in power do something about various social, economic, and environmental problems; my apparent freedom is that I do not have to assume responsibility for the problem at hand.

Technology can only continue to progress through constant experimentation and change. Through the images of the media, technology creates and manipulates desire. Public opinion is ephemeral, especially in its demand that consumer desires be fulfilled; for desire is always changing. Based as it is upon desire and fear, public opinion defines each issue in terms of individual well-being (happiness and health).

The visual images of the media are part of the ephemeral side of a technological civilization, the part that provides the greatest compensation for the demands that technology makes upon us. The media serve us dramatized but reverse images of a technological civilization, which is abstract and impersonal. The images of the media become as it were the language of technology. The visual images of the media are in harmony with public opinion in that they appear variously as an accurate representation of *what is* and as an imaginative alternative of *what is possible*. Technology's psychological hold over us is at the level of possibility.

Media images dramatize and make material each possibility, thus turning it into a model for action. In the 1970s a team of Canadian psychologists made a study of three communities in Western Canada, one of which was without television (because of its geographical location) but due to receive it within a year. The communities were studied both before television was introduced to the one community without it and two years afterwards. The study's main purpose was to ascertain the impact of television upon the attitudes, thought processes, and behavior of the residents. Most of the attention was devoted to children, but adults were studied as well. Television watching in Notel, the town previously without television, slowed down

33. Boorstin, *Democratic Experience*, 227–44.

34. Tocqueville, *Democracy in America*, 246–61.

children's "acquisition of fluent reading skills"; moreover, children who watched a lot of television were poorer readers than those who watched but little (correcting for intelligence). The introduction of television to Notel reduced the level of creativity among its children to what it was in the other two towns. Children's sex-role attitudes became markedly stereotyped after the introduction of television in Notel. Perhaps most importantly, television, significantly increased the amount of aggressive behavior, both verbal and physical, among children.[35] This, I think, is the definitive study of television images acting as models for behavior.

My freedom appears as the possibility of emulating celebrities (anyone who appears in the media) in their appearance, lifestyle, and behavior. Yet these visual images are collectivistic insofar as they objectify our experiences and choices. Visual images in the media present us with lifestyles, experiences, and commodities that are spectacular, and public opinion demands their realization. *This pseudo-morality replaces the dualism of the normal and the ideal (the is and the ought to be) with the dualism of the normal and the possible.* Our ideal is not transcendent now but one of human construction—a technological utopia.

The Role of "Morality" in Psychological Totalitarianism

This pseudo-morality plays a key role in the operation of an extreme collectivism, of what one might call psychological totalitarianism. Political totalitarianism was always founded on an ideology that made the nation, the race, or the ethnic group sacred. That kind of totalitarianism was symbolically anchored. It was effective to the extent that the political symbolism was accepted by the masses. The psychological pressures of political totalitarianism depended upon shared political meaning however terrible it might be.

The effectiveness of the mass media, however, as the key agent of psychological totalitarianism is not based on political or religious ideology. Rather it rests upon a base that I have described elsewhere as the myth of technological utopianism. Unlike religious myths in which meaning was spiritual—nature or the gods—this myth is thoroughly materialistic. Technological utopianism substitutes the perfect health and happiness of the human body for the spiritual well-being of the human soul. This meaning is ineffective because it is based on individualistic consumerism. For meaning to be effective it must be shared meaning that binds people together in common responsibilities and reciprocal moral relationships.

35. Williams, *Impact of Television.*

Consumerism is a shared belief but it leaves one psychologically isolated, for it is based upon freedom without responsibility. The attempt to create meaning in consumerism, to spiritualize consumerism, fails because its utopian promise of perfect happiness and health cannot be achieved in this world, and therefore happiness and health remain transitory, as anxiety, suffering, and death constantly remind us.

Moreover, the reciprocal moral relationships that work to create trust have been replaced by obligations to conform that are a result of psychological weakness or the fear of others. Public opinion and the images of the media serve as a substitute morality, then, not out of shared responsibility and shared meaning, but out of the mutual fear of isolated and anxious individuals. We conform to the "moral" pressures of public opinion, the peer group, and the images of the media as well because we have come to believe that what public opinion desires and what the images of the media dramatize and make material are just what we desire as individuals. Our individuality becomes a random collection of the accidental differences that ensue from consumer choice, so that the collectivism of forced consumption remains hidden. This pseudo-morality provides the binding force of psychological totalitarianism. Psychological totalitarianism and its morality function according to *individual desire, not shared meaning.*

An Almost Impenetrable Moral Ambiguity

Everywhere there is evidence that repudiates my theory. The widespread discussion about ethics today seems to indicate that traditional morality is more or less intact. Every organization, every professional group, every political assembly, every corporation, establishes a code of ethics. Ethicists are in demand in medical schools, business schools, and law schools. Yet despite school ethics programs, parents and teachers are worried about the ethics of children and young adults. For example, student approval of cheating under certain circumstances is quite high.[36] The concern for ethics has perhaps an air of desperation about it. There is, I think, a tacit recognition that a common morality has disappeared, or at least that we have not worked out a moral response to the numerous problems with which technology in particular confronts us, e.g., cloning, genetic engineering, global warming, and so forth. The reason is our failure to recognize the true context of these problems. In *Medical Power and Medical Ethics,*[37] J. H. van den Berg, himself a physician, argues that, when medicine was less technologically effective, it

36. Hunter, *Death of Character*, 160–65.
37. van den Berg, *Medical Power*.

did everything it could to prolong life. But today, with the greater power of medicine, the extension of life sometimes leads to enormous suffering for the patient and her family. The technological context of the attempt to keep patients alive creates a moral problem of prolonged suffering that coexists with increased life expectancy.

At other times the concern for ethics appears as ideological compensation for the absence of moral restraints. One study in the United States discovered no relationship between business ethics and business practices. Indeed, Robert Jackall's *Moral Mazes*[38] confirms this finding about corporations.

I do not wish to impugn the motivation of well-intentioned citizens who seek moral solutions to the multitude of problems we face today, but for those in business or government who use ethics as a cover for cynical practices we should have little sympathy. The first confusion about morality today is that between theoretical morality and lived morality. Theoretical morality or normative ethics is a rational account of the virtues and principles people should live by. It is sometimes the work of a philosopher or religious leader and is based on revelation or some idea of natural law. Ellul[39] observes that historically there are only a relatively few instances in which the theoretical morality (normative ethics) has actually become the lived morality or effective moral attitudes of a society. Christianity is one example. Even here the time of agreement was rather short. The lived morality rather quickly departs from its ideal formulation. The main reason is that lived morality is most often a spontaneous, largely unconscious creation of a society, reflecting in contradictory fashion both *necessity* and an *ideal image* a group has of itself.

Some of the discussion of ethics today is at the level of theoretical morality and, as previously discussed, tends to take an issue out of its proper context—a technological context—or is ideological compensation for the disappearance of a traditional morality. Hence much of the current discussion of morality is beside the point, as our earlier discussion of the teaching of ethics in school indicated.

The second confusion about morality is between moral custom and lived morality. Moral custom is the lived morality of the past that survives into the present. It is only effective when it does not contradict the dominant lived morality of the present. Today both Judaic-Christian and humanistic moralities have been relegated to the status of moral custom. They still operate at times in personal relationships not fully subject to the technical and

38. Jackall, *Moral Mazes.*
39. Ellul, *To Will.*

aesthetical norms of the pseudo morality. Even then, it takes almost heroic strength and acute insight to live within the fading image of moral custom.

Let us take parental love of children as an example. Aesthetical love—that is, love that is based upon attraction and pleasure, and that is interesting and immediately fulfilling—has gradually supplanted ethical love based upon a sense of obligation to and even sacrifice of one's interests for the other. Martha Wolfenstein analyzed American governmental brochures provided to new parents over a number of decades. She discovered that later documents much more than the earlier ones emphasized the pleasure the child brought the mother. Wolfenstein discusses this in the context of what she terms fun morality: We have an obligation to have fun.[40] The second example is from Beatrice Gottlieb's analysis of medieval parental love. She observes that love was an element of moral discipline for Christian parents, one of whose chief obligations was the formation of the child's character.[41] Much later in the nineteenth and especially twentieth centuries, love and discipline became separated to a great extent so that love became the giving or receiving of affection, and discipline became a punitive form of control. Parents may need to employ both love and discipline, but the two have now become distinct so that love could be aestheticized.

The third example of moral confusion is the tendency of modern parents to live vicariously through their children's accomplishments, to need their children's affection too desperately, and to wish to become friends with their children before they are even adults. The child is thus turned into an emotional commodity. Despite the fact that parental love is markedly different today from what it was in the Middle Ages, the same word *love* is used in both instances. Therefore parents may readily miss the transformation of ethical love into a sentimental love that is primarily if not exclusively an aesthetic experience.

Another problem parents face in living out traditional morality in relation to their children is the diminution of parental responsibility. The state and human services experts have taken over many of the parents' responsibilities. Governmental regulations about child-rearing and services for families as well as expert advice on child-rearing reduce parental responsibility to the level of making the best consumer choice on behalf of the child. This is minimal moral responsibility.

My point is that the confusion between moral custom and lived morality and between theoretical morality and lived morality means that many will

---

40. Wolfenstein, "Emergence of Fun Morality."
41. Gottlieb, *Family in the Western World.*

remain unaware that traditional morality is rapidly disappearing. Current discussion of morality only compounds the problem.

Emile Durkheim was correct about normlessness or what he called "anomie" as a condition of modern life, although it is clear now that anomie was in its infancy when he made this observation. But he was wrong about it in a way that he could never have anticipated: an anti-morality would replace traditional morality. And it is this pseudo-morality that reinforces the meaninglessness of modern life. As this pseudo-morality radically individualizes us, technology and its ally global capitalism can more easily bring us under their dominion.

Chapter 8 _____

Visual Morality and Tradition:
Society's New Norms

Originally published in the *Bulletin of Science, Technology and Society* 20, no. 1 (2000) 29–34.

F rench sociologist Jacques Ellul once remarked that visual images are "norms in a world without meaning."[1] By contrast, traditions, which form the context for moral norms in historical societies, are replete with meaning. Traditions, which are retained in collective memory through ceremonial observance and ritual, both symbolize and interpret important events in nature or in society, such as the creation of the world or the birth of a society. But technology, as Ellul powerfully argued, tends to supplant tradition, understood as the shared wisdom of the past.[2] The decline of tradition is accompanied by the inefficacy of symbols.

Technology is being used here to refer both to material technologies such as machines, and nonmaterial technologies such as advertising or administrative techniques. Until the advent of a technological civilization, religious and political symbols, which provided humans both with intellectual and psychological confidence and a sense of purpose, were a principal way of exercising control over nature. Today, symbols have become sterilized. Because technology is our own invention, we perceive no need to symbolize it in myth, literature, and history.[3] Modern symbols have been transformed into ephemeral, vague symbols that are imposed on us through psychological manipulation—political propaganda, advertising, and public relations. For example, both Volvo and Buick have offered an ad: "Buick (substitute Volvo), something to believe in." We are now expected to believe in

1. Ellul, *Humiliation of the Word*, 147.
2. Ellul, *Technological Society*.
3. Ellul, "Symbolic Function."

consumer goods as if each were a religious denomination. Shared meaning is dependent on efficacious symbols that are experienced as both true and long lasting. Hence, we live in a technological civilization that is eminently powerful and nearly meaningless. Let me explain.

Because technology is simply the most powerful means of acting, to say that today we believe in technology means to say that we have turned power into a value.[4] Genuine moral values, however, place some limitation on power. For instance, freedom necessitates that there be limitations on the exercise of political power, just as love requires that each individual limit their selfishness in relating to others. Power has become not only a pseudo-value but also a near-exclusive value.

Information is never neutral; it directs our behavior, including moral conduct. Now the information that remained in traditional societies was practical knowledge useful for the group's survival and given symbolic meaning. Information about hunting, for example, was both about traditional ways of hunting and how the deities or their representative, the "master of the hunt," provided the animals, whom humans must respect, even as they hunted them. Practical knowledge thus was ritualized because it was set within a narrative about the meaning of hunting.

Today, information tends to go in opposite but mutually reinforcing directions. Knowledge has become abstract, both as theoretical knowledge and technical knowledge.[5] Technical knowledge tends to supplant tradition or common experience in providing information that is practical in the sense of efficient, objective (decontextualized), and meaningless. Such abstract information is not sufficient to meet human emotional needs. Traditional information did this by providing a symbolic context for practical knowledge. As shared meaning declines in a technological civilization, knowledge must develop that is emotional as an end in itself rather than as an epiphenomenon of shared meaning. Emotional knowledge becomes, then, a compensation for abstract knowledge, especially technical knowledge. And the most prevalent source of emotional information today is the televised visual image. Visual images register first and foremost an emotional impact on the viewer.[6] By contrast, language, which certainly possesses an emotional dimension, makes a fundamental appeal to the mind unless it is used for purposes of psychological manipulation. I want to consider how the dramatized

4. Ellul, *Technological Bluff*, 109.

5. Ellul, *Technological Bluff*, 324–46.

6. Gombrich, "Visual Image."

information of television creates "moral norms." To do so, we first must reflect on the concepts of truth and reality.[7]

Ultimate meaning (the meaning of life, the meaning of history) is related to the order of truth. By this is meant the truth as a cultural form without specifying its content. We thus can speak about truth in a general way. The order of truth refers to the relationship between my actual words or actions and their ethical counterpart, what they should be. Moreover, it refers to the relationship between my words and my actions. Is my life true or false in regard to what I profess to believe? Truth and falsehood both belong to the order to truth. That is, as dialectical concepts, each term implies the other and is defined in relation to its opposite. The order of truth suggests an ethical ideal (even if not fully knowable) according to which words and actions can either be true or false. By contrast, the order of reality refers simply to whether something in the empirical world exists. Factual truth is a mixed case. On one hand, it refers to whether what I say corresponds to what is, but on the other hand, it refers to an ethical ideal that there should be correspondence between words and reality. With the triumph of modern science, ethical truth in the traditional sense of the term increasingly is perceived as subjective; therefore, it loses much of its efficacy.

Only discourse allows us to express our experiences of the truth. Discourse can be used to describe reality, but its genius is to help us pursue the truth. By itself, the visual image refers us to an empirical reality, actual or imaginary. Although the visual image and discourse both can be used to describe reality, the image is a more compelling source of empirical information.

Now what many have called reality, at least until recently, is actually a dialectic of the two orders of empirical reality and truth. Because societies more or less follow a common morality that places some limitation on power and retains some memory of the truth, reality is experienced as more than empirical reality. A sense of the mystery of life appears universal. What makes life mysterious, of course, is not just its incomprehensibility but the intimations of something transcendent.

A reality that is a union of the two orders of empirical reality and truth begins to wane with the triumph of science and technology. If modern science provides the intellectual superstructure for a thoroughgoing materialistic worldview, the visual images of the mass media supply its emotional infrastructure.

Television is the mass medium that most dominates our lives. The average American household has the television on fifty-five hours per week; the

7. Stivers, *Culture of Cynicism*.

average individual, depending on age, watches television from a little more than three hours a day to more than five hours a day. For many, it is the single most frequent and important leisure activity. The following discussion is mainly directed toward television, although it applies at times to other visual media as well. I shall attempt to make a strong case that the visual images of the media, although invariably accompanied by the spoken or written word, dominate the form and content of what is communicated. The images of television in particular have come to constitute a kind of language.

Television appears to the viewer to be a description of reality; moreover, reality is perceived to be on television. These two propositions—television is about reality, and reality is on television—are different but complementary ideas. There appears to be a one-to-one relationship between visual images and reality (actual or imagined). As has already been noted, the visual image is related to the order of reality. Sight is the sense that we most trust. Images are understood to be operational indicators of what they represent. The image and reality create a unity that allows us certainty about the empirical world. The image provides us with the information necessary to manipulate and adjust to reality.

Television appears to be describing reality, particularly in news programs, documentaries, talk shows, and game shows. In effect, it is reconstructing reality by taking reality out of its temporal and cultural context. Reality as we live it still retains, no matter how small, some meaning (order of truth); but television expunges this meaning and recomposes reality as a logical sequence of image fragments. Reality appears in televised images as a series of objects that encounter one another in space.

There are many other programs on television, however, such as soap operas, prime-time dramas, and comedies that appear cognitively to be unreal. But emotionally, they are real, just as real as the programs that appear more congruous with empirical reality. At an emotional level, all images are real: all images register an impact on the emotions.

Jerry Mander, in reflecting on his own child's inability to distinguish television from reality, remarks that "*Seeing things on television as false and unreal is learned.*"[8] Just as with any other type of learning, this one may prove unsuccessful in certain cases. Gerbner and Gross's study of television viewing indicates that heavy watchers had a greatly distorted sense of reality.[9] Among other things, heavy viewers "overestimated the percentage of the population who have professional jobs; and they drastically overestimated the number of police in the U.S. and the amount of violence." The errors in the heavy

8. Mander, *Elimination of Television.*
9. Gerbner and Gross, "Living with Television."

viewers' judgments were congruent with the distortion of television. It is hard to escape the conclusion that "the more television people watched, the more their view of the world matched television reality."

Edmund Carpenter reports on an informal study of college students, who when told that television's report of an event was inaccurate, either accepted the inconsistency uncritically or doubted reality.[10] Television provides a more emotionally satisfying reality.

We have now reached the turning point: not only is television about reality but reality is on television. This assertion suggests that as spectators, we live vicariously through and in visual images and that the medium of television is a principal epistemological authority that rivals the computer in terms of efficacy. When television momentarily focuses on an issue, it becomes a "real" issue; when television leaves it behind for a new issue, the old issue is eliminated from our immediate experiences and thus becomes a nonissue. Even the trial of O. J. Simpson, perhaps the longest playing spectacle in recent memory, is for all intents and purposes a nonissue, even with the onset of the civil trial. A society of the spectacle turns life into a visual drama, a show; and the most real show, the most "objective" show, is that on television. Television has become the most important authority of dramatized information among the various mass media. We are fascinated with the possibility of being on television or of meeting someone who is on television (a celebrity), because it is only here, on television, that my subjective experiences (which are experienced emotionally as real) can be validated as real.

The visual images of television (and the mass media) are only images of power and possessions. Leaving aside dialogue for the moment, autonomous visual images are about objects (possessions) and their relations (power), or in other words, how some objects bring about the movement of other objects. Television makes all objects more or less equal because visual images differ only quantitatively in their believability and their potential for emotional arousal; moreover, in a consumption-oriented society, all objects become equally consumable. The difference between human beings and products disappears: human beings are constantly being interrupted on television by advertised products and, more important, have themselves become reified products.

The relations between products are relations of power: possession, consumption, manipulation, control, and violence. A world devoid of symbolic meaning in the sense of ethical truth is a world of raw power. Because symbolic meaning flows directly or indirectly from discourse, autonomous

10. Carpenter, *Oh, What a Blow*, 61–66.

visual images are only about relations of power. For example, on television, a man and a woman embrace. Is this an instance of ethical love, sexual desire, or the desire to manipulate the other? The visual images by themselves make the first response impossible, because the visual image is about the order of empirical reality. We are left with a choice between or a combination of the second and third responses.

But even when we factor in discourse, the visual images of the mass media are still only images of possessions and power. The subversion of language by the visual images of the media, particularly television, is the result of the convergence of three phenomena: (a) the logical linking together of visual images to create a pseudo-language, (b) the visual images of the media as a more emotionally satisfying reality than the one we experience directly, and (c) the impoverishment of natural language.

The visual images of television cut up and reconstruct reality according to a spatial logic of signifiers. The "meaning" of a television program is the final outcome—success or failure, survival or death, possession or nonpossession, in other words, power. The content of the various programs are about objects to be possessed and consumed, the power of which becomes that of the spectator/consumer; more exactly, the content is about objects acting in relation to other objects—acts of possession, consumption, manipulation, control, or violence. These acts of power are spatially linked together in a television program or movie in a logical sequence that leads to an outcome—success or failure.

Television provides us with a reality at once both spectacular and pleasurable; however, it is a meaningless reality. At worst, this reality is boring. Existence is, as we have seen, a symbolic reality (dialectic of the order of truth and order of reality). Reality, as we live it, involves suffering and is ultimately tragic. Only a sense of ultimate meaning, ethical truth, can allow one to face this reality honestly and begin to transcend it. Otherwise, this reality is unbearable; experienced as such, one is driven to escape its hold. As our lived reality becomes less meaningful and thus more terrifying, television and other media comfort us in the warm prison house of a spectacular reality.

The decline of natural language is signaled by several related changes in discourse: (a) rapid change in the meaning of symbols, (b) vagueness of meaning, and (c) reification of symbols. Recent studies point to opposite trends in discourse. An increasing number of words have become technical terms with one operationalized meaning; at the same time, a large number of words have become both abstract and vague. The latter evoke an emotional rather than an intellectual reaction. The two tendencies, a greater rationalization of discourse and a greater emphasis on emotion,

perfectly reflect a technological civilization, which simultaneously en-
larges the sphere of technical reason and that of instinctual emotion at the
expense of normative reason.

The decline of discourse is marked by the proliferation of symbols that
are ephemeral. Advertising turns every product into a symbol. Cologne
symbolizes sexual prowess, the sports car, success, and so forth. But these
symbols have no staying power, because advertising must constantly experiment with pseudo-meaning. The products come and go, and so do their
symbolic meanings. These symbols are "sterilized symbols," because they
(not the products but the symbolic meanings) no longer have a practical
purpose in a technological civilization as an integral part of culture; rather,
they serve as a means of propaganda. Political symbols suffer the same fate
as advertising symbols. Politicians are sold like products.

A key issue in regard to modern discourse is its vagueness of meaning.
By this, I mean a loosening of the connection between sense and referent.
The meaning of an utterance involves both sense (the what is said) and referent (the about what) of discourse. Vagueness in meaning can occur in two
ways. First, the same word, sentence, or symbol can be used in relation to
too many referents. The word *democracy* is used to refer to virtually every
extant government, no matter how totalitarian it is, by one party or another.
Second, the word or symbol may come to possess no referents. Qualitative
concepts such as love and freedom may increasingly have nothing to refer
to in lived reality. As has been argued, a technological civilization makes
ethical meaning superfluous.

Finally, the meaning of many key concepts has been reified: visual
images become the operational indicators of qualitative concepts. At the
same time, the symbolic meaning becomes vague and thus subjective;
it likewise becomes objectified in visual images. Language is becoming
subordinate to visual images. This is perhaps less serious with regard to
positive concepts, that is, concepts that refer to objects; but in respect to
dialectical concepts, that is, concepts that refer to qualities, it is devastating. The turning of language into an accessory to the visual image suggests that every concept expressed in discourse refers to an "object." The
result is thoroughgoing quantitative reality. When the protagonist says "I
love you" to the object of his affection in the soap opera, the "meaning"
of love, is the passionate embrace that accompanies the dialogue. There is
evidence, moreover, that an increasing number of people, especially the
heaviest television viewers, are losing their ability to think abstractly in
and through language. Some now think in largely emotional terms; that is,
they think from image to image (or emotion to emotion).

The subjectivization (sterilization) of symbols and their reification in visual images effectively reduces meaning to instinctual power. Visual images hit us at an emotional level. When visual images are subordinate to language and symbolic meaning, as in traditional art, the emotions unleashed are integrated by normative reason and made meaningful. When, on the other hand, visual images become autonomous, reified symbols, they leave the emotions under the control of the instincts: survival, aggression, sexuality, and so forth. For the individual (a spectacular reality creates a radical individualism), reality is emotional and meaning is instinctual. The implications of this are astounding. Technology is first and foremost an efficient or powerful means of acting, the visual images are images of power and possession, and the "meaning" of autonomous visual images is instinctual power. The circle is now complete: a reality of power, a reality without meaning.

Even when one allows for the discourse accompanying the visual images of television and the movies to provide meaning for the action, the primary mode of human interaction is domination/submission. In commenting on several media, Andrew Tudor argues, "Violent actions themselves are only the logical extension of this basically coercive image of human relations."[11] Television as a medium favors the more spectacular, peak events such as catastrophes, war, violence, death, sexuality, and conflict of any kind. This is the bias of the medium whose sole purpose is to entertain us. The more subtle forms of human interaction such as compassion, ethical love, trust, and patience cannot be transmitted through autonomous visual images; and even if they could be, they are not as interesting. The expression of power is more spectacular than the limitation of power.

Visual images make a fundamental appeal to our emotions. Visual reality is an emotional reality. At this level, the intellectual credibility of the visual image is less important than how spectacular it is, the emotional response it can produce. Studies indicate that the most believable (emotionally speaking) of realistic depictions on television are violent scenes. Violent images are the most real images because they provoke the strongest emotional response: They simultaneously give us a sense of being alive and of having control over our relations with others. In an existence increasingly made abstract, impersonal, and meaningless by technology, unusual, spectacular, and frightening images allow us vicariously to experience a crisis, a turning point in our lives. We are placed time and again in a crucible.

The violence on television and in movies often functions as a kind of magic. Violence solves human problems; it can even put an end to violence. From the martial arts films to those of Clint Eastwood and Arnold

11. Tudor, *Image and Influence*, 215.

Schwarzenegger, the good guy is the one who is ultimately most violent. Power and virtue are equivalent; moreover, the more powerful the action, the more virtuous it is. Violence as magic is the mirror image of technology. We look to technology as the solution to all problems. Violence in the media is a reverse image of technology. Technology is abstract and rational; violence is concrete and irrational. Both are expressions of ultimate power.

The compelling nature of images of power (coercion and violence) is a result of the conjunction of three factors: a technological reality, which is itself coercive and manipulative; the mass media, which overrepresents the amount of physical violence; autonomous visual images, which are exclusively images of power. Both the form and content of television reduce reality to a struggle for power. And the reality beyond television has come to reflect its own image.

The extent to which television motivates one to act (consumption not included), rather than merely live vicariously, is another matter. Natural experiments in the social sciences are vastly superior to laboratory experiments for most issues because artificial experiments take what one is studying out of context. In the 1970s, a team of Canadian psychologists made a study of three communities in Western Canada, one of which was without television (because of its geographical location) but due to receive it within a year. The communities were studied, both before television was introduced to the one community without it and two years afterward. The study's main purpose was to ascertain the impact of television on the attitudes, thought processes, and behavior of the residents of Notel (a pseudonym). Most of the attention was devoted to children, but adults were studied as well. The following conclusions are only a partial list of what they discovered.

Television watching in Notel slowed down children's "acquisition of fluent reading skills"; moreover, children who watched a lot of television were poorer readers than those who watched little (correcting for intelligence). The introduction of television to Notel reduced the level of creativity among its children to what it was in the other two towns. Children's sex-role attitudes became markedly stereotyped after the introduction of television in Notel.[12]

Most important, for our purposes, are their findings about television's impact on the aggressive behavior of children. The researchers classified aggression according to physical and verbal behavior. The physical behaviors included, among others, hitting, slapping, punching, biting, spitting, snatching property, damaging property, threatening with a held object, and so forth. The verbal behaviors numbered humiliation, disparagement,

12. Williams, *Impact of Television*, 395–401.

condemnation, rejection, and threat. Their conclusions, because they are so crucial, are worth quoting.

> The aggressive behavior of Notel children increased significantly following the introduction of television. This conclusion that TV viewing and aggression are linked has been reached by most other researchers, but several aspects of our results are new. They are based on observations of actual behavior rather than self-reports or ratings. Effects occurred for both girls and boys and for both physical and verbal aggression. Increases occurred for children initially low in aggression, not just a small subsample of highly aggressive children. The effects were substantial enough to be observable two years after the introduction of television to Notel.[13]

The autonomous visual image blurs the distinction between truth and reality. Factual truth becomes the perceived correspondence between image and reality. The order of truth is reduced to the order of reality in a culture of radical immanentism. Empirical reality is self-contained in this view; there is no sense of transcendent purpose. Reality by itself presents us with two possibilities: adjustment and manipulation. The latter is more advantageous to one's interests than the former, because it is an expression of power over reality, the power to create reality. Truth becomes, then, in its pseudo-ethical form success relative to reality. The implications of this are that both technology and the visual images of the media allow us to simulate reality and to create reality.

There is pause for reflection here. In the traditional conception of truth, factual truth is guided by normative truth (derived from a religious or philosophical conception of the good). In the modern conception, factual truth is based on a scientific worldview; technology as the most powerful means of creating or reconstructing material reality acts as its pseudo-norm. In the traditional view, reality is to be brought as nearly as possible into correspondence with that that is perceived to be an absolute, transcendent good. By contrast, the modern view is that reality should be molded by and reflect the possibility of technology. Normative truth is now the power to make reality. Normative truth was always a source of hope to the extent that normative truth, it was believed, could come to play a large part in lived reality.

The contrast between factual truth and normative truth in televised visual images is best represented by the news and advertising. As part of television, both the news and advertising present us with an ahistorical, fragmented world. The former concentrates on problems, many of which

13. Williams, *Impact of Television*, 401.

defy solution; the latter, however, provides an abundance of solutions. As Neil Postman observed, the typical commercial argues either or both of the following: (a) the technological object will solve a problem for the consumer, for example, how to be popular, how to be successful; (b) the technological object will make the consumer immeasurably happy.[14] The news is to advertising as factual truth is to normative truth (the ability to create reality). The mythological world of advertising solves problems that in the world of current events seem hopeless. Technology is thus that that is both true and good.

The visual images of the mass media are the "language" of technology and a chief form of its "morality": norms of the social reality technology has already created and norms of what technology makes possible. The visual images of the media are anti-traditions, ephemeral but coercive norms that bind us to a technological reality.

14. Postman, "Parable of the Ring," 66–81.

Chapter 9 ————————————————————

Sin as Addiction in Our "Brave New World"

Originally published in *Ellul Forum* 59 (2017) 23–27.

W e know from Scripture that humans sin, are born in sin, and are in bondage to sin. Biblical ideas of sin have a hard time being recognized today, however. Liberal Christianity has de-emphasized sin or reduced it to injustice and inequality. Conservative Christianity has tended to equate sin with personal immorality. In either instance, the truth about sin has been diminished. As Søren Kierkegaard reminded us, sin is not merely a matter of discrete sins but of an orientation, a way of life. Furthermore, Scripture makes sin a spiritual matter, not just a moral issue. Idolatry is the worst sin.

To overcome sin we must contest various evil powers as well as our own desires. In *If You are the Son of God*, Jacques Ellul argues that one of the meanings of sin is that of an external power that influences or even controls us.[1] The evil powers that Scripture reveals to us do not have an independent existence; they exist only in and through their relations to us. But they are real! There is no principle of evil nor an evil god. In a sense, the evil powers are our unintended creation. Money and political power, for example, are evil powers. Money and politics are not evil in themselves but in the spiritual value we attribute to them.

Scripture indicates that sin is both individual and corporate. The very concept of the "world" suggests as much. Cultures are anchored by a sense of the sacred, that is, by that which is experienced as absolute power, reality, and meaning. Examples of the sacred include nature, the tribe, money, and the nation state. The socially constructed sacred (tacitly, not consciously) provides both meaning and the basis for control in society. All social institutions obtain cultural authority as a result. *Exousia* refers to a spiritual

---

1. Ellul, *Son of God*.

123

power that the social group employs beyond that which it receives from its cultural mandate. The social group thus becomes more than the sum of its parts, spiritually and not just psychologically. But *exousia* refers to a material power as well.

All members of the group are motivated by covetousness and the will to power, which are the source of sin. The social group provides an absolute identity for the individual and excites the individual's desire through its internal competition for wealth and power. Hence the group is held together in part by the negative unity of sin. Social institutions do not fully control the will to power, for the exercise of power invariably exceeds the limits that cultural authority imposes on it. This excessive power (*exousia*) is both material and spiritual, power and value, human and alien. Sin is, in turn, both internal and external, individual and collective.

Scripture is replete with figures of speech, especially metaphors. God, for instance, is king, fortress, shepherd, and so forth. A metaphor is not to be taken literally, of course; it entails a comparison. What is less well known is compared to what is better known: God is compared to a fortress. No one metaphor is sufficient, for each metaphor reveals different aspects of the phenomenon. To say that "love is a rose" suggests that love blooms and fades, is fragrant, and is capable of inflicting pain. "Love is a journey" implies that love is not static and that the movement may be more important than the final destination. Unlike the logical concept, metaphor never permits us to pretend to grasp the phenomenon as it is in itself. The numerous metaphors about God are a warning not to claim to define and know God as he is. We apprehend God by comparison.

Often neglected in discussions of metaphor is the status of the better-known term. For metaphor to be vital, the better-known term must be common. The metaphorical comparison necessitates reflection on both terms. Consequently, we learn more about what we ordinarily take for granted, the better-known term. This will become apparent as we examine the following metaphors of sin.

The most prevalent metaphor for sin in Scripture is sin is bondage or slavery. John, Paul, and Peter refer to sin this way. Jesus says, "Everyone who commits sin is a slave to sin" (John 8:34). Paul states, "For freedom Christ has set us free; stand fast therefore and do not submit again to a yoke of slavery" (Gal 5:1). Peter proclaims, "They promise them freedom, but they themselves are slaves of corruption; for whatever overcomes a man, to that he is enslaved" (2 Pet 2:19). Slavery was widespread in the Roman world, and it was well understood that it takes away secular freedom. In attempting to understand sin, which destroys Christian freedom, the early Christians employed the metaphor that "sin is slavery." In doing so, they make us reflect on the *institution* of slavery.

In *The Ethics of Freedom*, Jacques Ellul suggests that "sin is alienation" is the metaphor that best resonates with our experiences today.[2] Ellul was not a Marxist, but he nonetheless employed Marx's concept of alienation. Under industrialized capitalism, the worker was alienated from his work, that is, he lost ownership and control over the process of work and the product. His work became merely a means of profit for the capitalist, who had made him a "wage slave." Because work was central to Marx's view of the human being, self-alienation followed alienation from work. To be alienated means to be possessed by another. Ellul's book was published in 1975, and parts of it were written in the 1960s. He understood that technology had become a more important factor than capitalism in the organization of society. Consequently, he applied the concept of alienation in a new way to demonstrate that in replacing human experience with objectified expertise, technology was itself alienating.

I think that today, however, another metaphor is more appropriate: "sin is addiction." Before examining addiction as a metaphor for sin, I should point out that all three metaphors, enslavement, alienation, and addiction, suggest being possessed by a person or force. Karl Barth once said that rather than say, "I have faith," I should say, "Faith has me." The three metaphors for sin suggest that I should say, "Sin has me," rather than, "I sin." In addition, all three metaphors reveal something about the larger society. To be enslaved makes manifest the institution of slavery; to be alienated reveals the institution of industrialized capitalism; to be addicted uncovers the technological system.

I will not attempt to define addiction in scientific terms. Is it physical, psychological, or both? Are there degrees of addiction? Instead, I will employ the term in its colloquial sense: something we can't seem to stop doing even though it's not necessary for our survival. Or a compulsion from which we can't or don't want to escape. Most people associate addiction with drugs and alcohol. Increasing numbers of people talk about addiction to social media, but the list of addictions keeps growing.

Julian Taber, who is a therapist to gambling addicts, developed the Consumer Lifestyle Index/Appetite Inventory.[3] It attempts to be a comprehensive list of addictions. The range of addictions is enormous: gambling for money, lying, laxatives, shopping, petty theft, sugar-based foods, tobacco products, exercise, talking for talking's sake, religious activity, work for the

2. Ellul, *Ethics of Freedom*.
3. Cited in Schüll, *Addiction by Design*, 242–43.

sake of being busy, trying to get attention for its own sake, self-help groups, and so forth. The obvious conclusion is that anything can become addictive. In "The Acceleration of Addictiveness," Paul Graham argues that technological progress brings more addictiveness.[4] Technological progress creates ever more products and services to which we may become addicted. Addiction to technology is the necessary result of technological progress. My point is not that addiction is omnipresent but that more of us are perceiving it this way. Talk of addiction brings in more conversationalists every day.

I will discuss addictions to machine gambling, video games, and social media in order to examine the metaphor that sin is addiction. We spend more money on casino gambling than on music, movies, and sports events together. Most of the gambling occurs with slot machines and video poker. One hundred and fifty-five million Americans play video games and spend more than twice as much on them as they do on movie tickets. Soon virtually everyone will have a smartphone or similar device to use Facebook, Twitter, Instagram, and other social media. Not all players and users are addicts, but much has already been written about the heavy use of these technologies as if it were an addiction.

Enslavement, alienation, and addiction all have sociological contexts. In the former, the context is an institution, in the latter, an entire social environment—technology. Following Jacques Ellul, by "technology" I mean both machines and nonmaterial technologies such as bureaucracy, advertising, and propaganda. Beginning in the eighteenth century, material and nonmaterial technologies advanced together. Nature and human society were increasingly brought under technology's purview. With the advent of the computer, it became possible to coordinate major technologies to form a system at the level of information. Technology has thus become a system. Human society now opens to two environments, nature and technology. Modern technology shattered the unity of culture. Technology supplants experience and meaning; it is solely about the most efficient (powerful) means of acting. Society is organized at the level of technology but disorganized at cultural and psychological levels. Culture is randomly created and fragmented in its meaning and purpose as a creation. The result is a plethora of moralities and art and entertainment styles. The lack of cultural unity makes psychological fragmentation inevitable: we are reduced to being role players who create multiple images for ourselves and others.

Technological growth has been accelerating for more than 150 years, although not evenly across the various sectors. Moreover, there appears to be no purpose or end to it. Implicit in the growth of technology is the

4. Graham, "Acceleration of Addictiveness."

mandate "If it can be done, it must be done." The traditional tension be-tween what is and what ought to be has been superseded by that between what is and what is possible. Consequently we have only limited moral control over the employment of technology. We have become as fatalistic about technology as so-called "primitive" people were about nature. Hence we have an irrational faith in technology.

Technology has an impact on the individual's psyche just as great as its influence on culture. Technology directly and indirectly provokes a need for ecstasy. The very point of addiction is to create a continuous ecstatic state. Ecstasy is an altered state of consciousness, an escape from the rational self. Ecstasy is a kind of high that can be achieved by rapid, repetitive movement, continuous loud music, drugs, and alcohol, for example.

Cultural anthropologists have a category of religion they call "ecstatic religion." It includes rites organized to produce an ecstatic state in the par-ticipants. Such rites may involve orgies, drunkenness, and violence. Victor Turner maintains that the rites designed for ecstasy bring about a communion of equals, a *communitas*, whereby status differences and power relationships are temporarily set aside.[5] A feeling results of one in all and all in one. Some have extended the meaning of ecstatic communion to include communion with machines. Today we have technology to help us achieve ecstasy.

Technological progress has increased the pace of life: we do more in less time. Speed has become an end in itself. Time urgency entails a compul-sion to do as many things as rapidly as possible, including a preoccupation with time, rushed speech and eating, driving too fast and angrily, waiting impatiently, and feeling irritable and bored when inactive. Concurrently, we suffer from time scarcity. Family life and leisure mimic the speed of the workplace. With mother and father both working and the children in a plethora of organized activities, parents have to become efficiency experts. Tourism and vacations typically involve stuffing as many activities as pos-sible into the shortest period of time.

Speed itself can produce a mild ecstatic experience. Milan Kundera observes that "speed is the form of ecstasy the technical revolution has be-stowed on man."[6] We internalize technological stimuli. Wolfgang Schivel-busch refers to this as the "stimulus shield."[7] We adjust to and normalize the ways that technology alters our sense of time, place, speed, sight, and sound. Each time a faster mode of transportation was introduced, people had to adjust to it, and eventually the previous mode seemed hopelessly

5. Turner, *Ritual Process*.
6. Kundera, *Slowness*, 2.
7. Schivelbusch, *Railway Journey*.

slow. Humans internalized the speed of the train, for example, and later, when given a choice, they rejected the horse and buggy. Today we internalize the speed of faster computers and are impatient when forced to use slower ones. We come to resemble the faster technology that stimulates us: we act by reflex, not reflection.

Technology creates a need for ecstasy as an escape mechanism. Anthropologist Roger Caillois observed that the more extensive and intensive the social controls in a society, the more exaggerated the ecstatic response.[8] We cannot tolerate living in a social world that is too ordered. Never before have humans lived with so many rules—technical, bureaucratic, and legal. The proliferation of administrative laws, bureaucratic norms, and technical rules that accompany each new technology makes it impossible for anyone to be aware of them, let alone remember them. We feel the pressure to escape them in irrational ways: drugs, alcohol, sex, sports, gambling, and so forth. A Columbia University psychiatrist found that the harder college students (especially males) studied during the week, the more they felt the need to escape the rational order of obtaining good grades by giving themselves over to instinctual desire and temporarily losing their conscious selves.[9]

Technology indirectly produces loneliness from which an escape is necessary. Christian psychiatrist J. H. van den Berg demonstrated that the loss of a common morality beginning in the eighteenth century in the West resulted in human relationships becoming vague and dangerous.[10] A common morality in society meant that one could trust people even if one did not especially like them. The decline in trust makes everyone a potential enemy. Loneliness ensues. Van den Berg argues that loneliness is the nucleus of psychiatry, and that all psychiatric disorders are intertwined because all patients share the same existence. For many of us, loneliness does not result in a full-blown psychiatric disorder, but the number of Americans in therapy, self-help groups, and on drugs for depression is legion. Loneliness manifests itself in many ways, some of which conceal the loneliness. One of them is the need to talk incessantly, sometimes to anyone who will listen, about trivial matters. I can't be lonely if I am talking to people! With the advent of email and social media, we can be in communication with others anytime we feel the need. The result is the ecstasy of communication. The speed by which information is transmitted from person to person produces a mild ecstatic state.

8. Caillois, *Man and the Sacred*; see also Ellul, *Technological Society*, 387–427.

9. Hendin, *Age of Sensation*.

10. van den Berg, *Changing Nature of Man*.

If technology creates a need for ecstatic release, it also produces the means to achieve ecstasy. Machine gambling is a prime example. In *Addiction by Design*, Natasha Schüll interviews gambling addicts and discovers that what they most crave, even more than winning, is the "zone," in which "time, space, and social identity are suspended in the mechanical rhythm of a repeating process."[11] In other words, a state of ecstasy. Gamblers enter the zone when their actions and the functioning of the machine become indistinguishable. Schüll borrows the term "perfect contingency" to describe the sense that addicted gamblers have of a perfect alignment between their actions and the machine's response. They prefer "sameness, repetition, rhythm, and routine."[12] Slot machines and video poker are the most popular gambling formats. As gamblers develop a tolerance for the technology (stimulus shield), the games become faster and more complex. For instance, in video poker, Triple Play Draw Poker allows players to play three games at once and make three times as many bets. Triple Play has given way to Five Play, Ten Play, Fifty Play, and even Hundred Play Poker.

Video game addicts too desire to merge with the machine, to achieve communion with it. In *God in the Machine: Video Games as Spiritual Pursuit*, Liel Leibovitz, himself a video game player, describes how reflex replaces cognitive awareness the greater one's skill and mastery becomes. His experience is mainly with the Legend of Zelda. Repetition is the foundation of play, from the "ballet of thumbs" to returning to the same play section without stop and with little if any variation. The spiritual pursuit that Leibovitz claims is the deeper rationale for playing video games is ecstasy. If ecstatic religion is a legitimate category of religion, then video games are a subcategory. In defense of his interpretation, Leibovitz argues that video games teach one the joy of learning to love the game and designer above all, of giving up "all other ways of being in the world" and of "understanding one's place in the world."[13] He calls this a kind of Augustinian condition. I am not arguing that his interpretation is correct but only that he points out how seriously we should take the pursuit of ecstasy through our technologies.

The social media are not ostensibly about communion with a machine but with other people. We must remember, however, that every technology that permits us to communicate with others mediates the relationship. Social media "addicts" appear to spend less time servicing their addiction than do gambling and video game addicts. Nonetheless, a large number of social media users admit that they cannot give up their devices, if only for a day.

---

11. Schüll, *Addiction by Design*, 1–27.
12. Schüll, *Addiction by Design*, 166–87.
13. Leibovitz, *God in the Machine*, 125.

In the smartphone industry, it is commonly thought that people check their phones at least 150 times a day. Some are even bedeviled by phantom ringing or vibrating phones. One third of Americans claim they would rather give up sex than their cell phones. But is this really about communion with others and creating a community?

In *Alone Together*, Sherry Turkle discovers that the community of one's friends, say, on Facebook, is both fragile and enslaving. On social media, people are role players, presenting a self to others that will be most accepted and admired. The relationships established in social media networks are purely aesthetical and superficial. Only face-to-face moral relationships are deep and truly passionate, Kierkegaard has noted. Indeed, the more time one spends on Facebook, the more lonely one feels. Turkle observes that many young people prefer texting someone to talking to her. The reason is that a call involves more commitment than a text. A call could prove unpleasant and demanding.[14]

The social media intensify the urge to conform to the group. Turkle discovered that some young people believe that everything they do in public will end up on Facebook or its equivalent. This leads to "anticipatory conformity." She also claims that the social media are producing "group feelings," or ecstatic communion.[15] Elias Canetti terms a group that becomes a unified whole the "open crowd," the truest expression of the crowd phenomenon.[16] Within the open crowd there is a sense of absolute equality, because all divisions among people are momentarily obliterated. The ecstasy that ensues from the use of the social media is not communion that establishes a community, but communion that creates an open crowd, always poised to become a mob. There is no freedom and love in the crowd. Because they wear the mask of love, the social media are the most pernicious of the addicting technologies.

Because we internalize technological stimuli (stimulus shield), we develop a tolerance for them and demand that they be even more intense. This is a classic problem in the acceleration of addiction. The technology industry is accommodating; it designs these technologies to be ever more addictive.

Those who design information and communication technologies and technological products design them to be addictive. In *Hooked*, Nir Eyal discusses in detail how to make products habit-forming.[17] The author has

14. Turkle, *Alone Together*.
15. Turkle, *Alone Together*, 262, 177.
16. Canetti, *Crowds and Power*, 16–17.
17. Eyal, *Hooked*.

a background in the video game industry and advertising and has taught courses on applied consumer psychology at the Stanford Graduate School of Business. His book is a manual on how to make technologies and products attractive and addictive. He makes no pretense that it is not about manipulating the consumer.

In his model of how to "hook" the consumer, the "trigger" is what sets the behavior in motion. "External triggers" contain information with directions about what to do next. Advertising and word of mouth can motivate the consumer to require a new app for her smartphone, for example. Eyal maintains that the key to creating addiction is the "internal trigger." Associating a product with desire or fear appears to be the supreme internal trigger. The strongest emotional triggers are visual images. Eyal mentions that the internal trigger for Facebook is the fear of missing out, and, for Instagram, the fear of losing a special moment. The design of variable rewards is essential. Research has indicated that the anticipation of a reward, rather than the reward itself, motivates users. One receives a reward on occasion but not constantly. Those cherished images of family and friends are received only intermittently.

In *Addiction by Design*, Natasha Schüll explores in great detail how the machine-gambling industry probes the psyche of the addict as an aid in designing gambling machines. Addicted gamblers want to play multiple hands or games as rapidly as possible without interruption. Variable rewards are built into the software of the machine to increase with the frequency of the smaller separate bets that gamblers prefer to make. Gamblers can thus enter "the zone" more quickly and stay there longer. Video game designers use a similar psychology to make their games more addictive.

We have entered a new phase of technological progress, in which there is a conscious effort to make us addicted to technology. This is nothing less than an intentional technological totalitarianism. Early on, we were only dimly aware of the totalitarian nature of the technological system. The technological system has now reached a stage in which experts openly discuss the desirability of the total psychological control of humans. Aldous Huxley's *Brave New World* almost perfectly anticipates today's technological totalitarianism. In his 1932 novel, Huxley talks about "conscription by consumption."[18] We are free, but only as consumers. In his dystopia, freedom is redefined as happiness. In this society, moral relationships are prohibited—no families or close friends—but only transitory, aesthetical ones. Perhaps his most brilliant insight was that pleasure was the chief agent of control. Sex, "soma" (an all-purpose drug for any psychological discomfort), and "the feelies" (cinema

18. Huxley, *Brave New World*, 33.

with full sensory stimuli) were the main obligatory pleasures. Huxley saw that group therapy would reinforce the controls technicians had established. Are we not in a brave new world with all our pharmaceuticals, self-help groups, social media, advertising, public relations, propaganda, experts on every aspect of life, culture reduced to its lowest level—entertainment—and widespread family dissolution?

What does addiction tell us about individual and corporate sin? Addiction takes possession to its zenith. Slavery and alienation both entail possession but not to the same extent. The metaphor of addiction demonstrates as well that pleasure is the key to sin's control over us. We love our sin. Addiction reveals the accelerating nature of sin: it is dynamic. We quickly sink deeper into sin. Finally, addiction reveals the totalitarian nature of sin. Sin wants all of us, all the time. These ideas are explicit or implicit in Scripture but not in the form of a single metaphor if only because addiction as we know it did not exist then.

Earlier I suggested that a metaphor makes us reflect on the better-known term, not just the lesser-known term. "Sin as slavery" tells us how the institution of slavery takes away our freedom or enslaves us. "Sin as alienation" informs us how industrialized capitalism strips away our freedom or alienates us. "Sin as addiction" instructs us about how the technological system eliminates our freedom or makes us addicts. Each metaphor invites us to reflect on the specific ways that the world, as the place of sin, controls us.

My point is not that gambling, playing video games, and using social media are evil in themselves but rather that *exousia* are at work in our social institutions with the intent of turning us into idolators. In our world, idolatry is best understood as addiction to technology.

Chapter 10 _____

The Festival in Light of the Theory of the Three Milieus: A Critique of Girard's Theory of Ritual Scapegoating

Originally published in the *Journal of the American Academy of Religion* LXI, no. 3 (1993) 505–38.

R ene Girard has challenged his critics to accept his hypothesis about mimetic desire and violent scapegoating as the origin of ritual and of culture itself or to propose an alternative explanation. This challenge is to be understood in the spirit of humility about his theory, for he is willing to enter into genuine dialogue about his seminal idea. I accept his challenge, not by rejecting his idea, but by modifying his hypothesis in light of an equally startling one—Ellul's theory of the three milieus.

Girard's hypothesis, that ritual scapegoating is a cultural solution to the contagious conflict engendered by mimetic desire, purports to be universal in regard to history and human nature, at least until the advent of an irreversible event. It is a solution that defies conscious criticism until the texts of the Old and New Testament expose ritual killing as scapegoating. Until that time human nature appears to be cast adrift in a torrent of mimetic desire only to be saved from universal spiraling violence through ritual acts of scapegoating. With the revelation of ritual killing as scapegoating, there is the opportunity for humans to confront both their violence and the violent solution to their violence.[1]

I maintain that Girard's theory fits only part of history and that the realization of ritual killing as scapegoating is due less to biblical revelation than to radical changes in history. In order to support my thesis I draw upon Jacques Ellul's theory of the three milieus.[2] Ellul's theory is a theory of

1. See Girard, *Violence and the Sacred*; Girard, *Scapegoat*; Girard, *Things Hidden*.
2. Ellul, *What I Believe*, 89–140.

history, but one that is *not* finalistic; it is dialectical at the level of existence. Unlike Girard, who posits a universal human nature, Ellul assumes one that changes as its life-milieu changes. The milieu, in part a human creation, becomes an objective force with which humans interact. This, I take it, is the origin of the sacred, the original act of reification.

I propose, then, to accept Girard's hypothesis in its entirety for the milieu of society but to reject it in respect to the earlier milieu of nature and the later milieu of technology. I argue that both desire and the sacred are experienced in relation to their various milieus and are best understood in the context of the ritualized festival, the meaning of which, in turn, changes according to its milieu. I begin with a summary of Girard's theory of mimetic desire and ritual scapegoating, proceed to a presentation of Ellul's theory of history, and ultimately place the two theories in juxtaposition to yield a theory of the sacred and of the ritualized festival.

Girard's Theory of Mimetic Desire and Ritual Scapegoating

Girard discovered mimetic desire, at least the tacit conception of such, in certain great works of literature—those of Shakespeare, Cervantes, Dostoevsky. Later he examined sacrificial rites in the context of religious myths and brilliantly articulated a theory of the relationship between mimetic desire and ritual sacrifice.

Mimetic desire is desire complicated by competition for the object of desire. Desire becomes mimetic when one desires what someone else desires; the other becomes a model for me: I desire to be like the person who desires what I now desire. I am thus attracted to this other person. Concurrently, however, this other is a rival: we both desire the same object. The object of desire grows in importance as the competition for it intensifies.

My rival gives two messages: be like me; do not be like me. The rival is flattered at first by imitation, for everyone wants to be admired; the initial feeling of pride is quickly countered by the fear of losing the object of common desire. As a rival, the other is an obstacle to the realization of desire. Hence, I am repelled by my rival. Attraction and repulsion.

Mimetic rivalry is contagious, as is the violence it engenders.[3] Violence is itself mimetic. I respond to your violent act with a similar act of my own. We become like each other in our violence. Still others enter the fray, for each momentary victory attracts new opponents. This is the mimetic crisis: the certain prospect of widespread violence with no apparent solution.

3. Dumochel, ed., "Introduction."

A solution emerges spontaneously, one that is the originary act from which ritual, religion, and society itself spring forth. The solution is a collective murder that both ends the violence and symbolically unites the remaining competitors—the killing of a common enemy, a *scapegoat*.[4] What may appear as an arbitrary act is not recognized as such by the collaborators; the scapegoat is not recognized as scapegoat. For the act of ritual killing to be effective as an agent of unification, it must not be recognized as an act of scapegoating. The scapegoat will be viewed as the source of all evil and thus as one who deserves to be killed.

In *The Scapegoat*, Girard explains the stereotypes of persecution by which a scapegoat is selected. The first stereotype is that of the crisis. Some event perceived as momentous, externally or internally provoked, causes people to act increasingly similarly, such as hoarding food in time of famine. Social order is invariably based upon structural and cultural differences that translate into different but complementary actions. Consequently, a great similarity in action is tantamount to social disorder. The second stereotype is about the "crime" that precipitated the homogeneity of behavior, the loss of social differences. The crime has a trail that leads to the perpetrator(s). The third stereotype concerns universal signs for the choice of the scapegoat. The victims are always those who are different, whether ethnically, culturally, or physically; the differences must be readily observable. The scapegoat is not consciously accused of being different but rather of "crimes that eliminate distinction." The eliminated distinctions are those between good and evil and between social ranks. The victim is ultimately condemned for *not* being different in a manner appropriate to cultural assumptions.

Here is the crux of the matter. As Girard explains, there are two kinds of differences—within and without the system. Diversity is a necessary component of social life: everyone has the need to appear different from others in some respects. Concurrently there must be a larger unity that envelops the various diversities and makes sense of them.[5]

By identifying a victim as the cause of the cause of disorder in society, as responsible for the original crisis (famine, plague, poisoning) that in turn leads to the crisis of contagious violent behavior, the scapegoaters are simultaneously establishing boundaries between the system and what lies outside and reestablishing diversity of action within the system. The scapegoat belongs outside the system because he caused the individuals within the system to shed their diversity and become dangerously alike.

4. Girard, *Scapegoat*.
5. Vanderburg, *Growth of Minds*, 27–29.

Girard argues that ritual scapegoating, sacrifice, and rites of all sorts have their origin in an act of killing. The murder of the scapegoat unites the members of the society by permitting evil to be expelled. Society cannot confront the arbitrary nature of its violence and of its own complicity in the contagious violence preceding the original act of scapegoating. Hence religious myths conceal the act of murder as scapegoating; instead they represent the murdered as fully deserving to be killed. Simultaneously, however, the ritual victim is sacralized, transformed into a deity who now can be made to work for the benefit of society. For instance, in the Seneca myth of creation, after the good twin (the Creator) defeats his brother the evil twin (the Great World Rim Dweller), a pact is made between the evil twin and the Seneca. In return for food and supplications, the evil twin would "cure incurable disease, avert deadly tornadoes, cast out malevolent witches, and bring order to a whole community."[6]

The originary act of murder is not just kept alive in myth; it is systematically repeated in ritual. The reenactment is often a dramatic enactment rather than an actual repetition of the originary act of killing. There may likewise be a substitution for the ritual victim—animals, even crops, may become symbolic surrogates. In Girard's view, then, the ambiguity of sacred value resides in this fundamental contradiction: violence (evil) is the origin of order (good). If violence is sacred, it is so only because it is the source of unity, the order of all against one.[7]

Ritual scapegoating in the modern era, however, does not deify the victim. Girard comments on this:

> Religious phenomena are essentially characterized by the double transference, the aggressive transference followed by the reconciliatory transference. The reconciliatory transference sacralized the victim and is the one most fragile, most easily lost, since to all evidence it does not occur until the mechanism has completely "played itself out." We remain capable, in other words, of hating our victims; we are no longer capable of worshipping them.[8]

The farther away from its violent origins a community and its religion get, the more ritual scapegoating loses its meaning and purpose. Eventually the point is reached where rituals that celebrate order and incorporation take precedence over rituals of disorder—those of transgression and

---

6. Wallace, *Death and Rebirth*, 92.

7. Girard, *Things Hidden*, 32.

8. Girard, *Things Hidden*, 37.

those of sacrifice.[9] The time comes when scapegoating will be justified for political reasons.[10]

Yet it is not the passage of time alone that desacralizes ritual scapegoating; rather it is the New Testament, the story of Christ, Girard argues.[11] In this nonsacrificial rendering of the New Testament, Girard attempts to demonstrate that Christ's purpose was not to become the perfect sacrificial victim, to atone for human sin, to make reparation to the Father; instead it was to overcome violence and to establish an existential paradigm of Christian love: to love one's enemy, even one's scapegoaters. In so doing, Christ exposed scapegoating as scapegoating, as an act of total duplicity. Whether the scapegoat was guilty of his purported crime is irrelevant—in Christ's case there was no crime of any kind—for the scapegoat has heaped upon his head the problems and sins of everyone. In the logic of scapegoating the crime of the one is the cause of the misery of the multitude. The New Testament (and the Old Testament to a lesser extent) permits us to establish the concept of the scapegoat. As Girard maintains, from this moment on, scapegoating can readily be understood for what it is. Its efficacy is now only transitory. Yet the violence is not abated because of this knowledge about the inner workings of scapegoating; it only becomes more diffuse. As Girard says at the close of *The Scapegoat*, to this knowledge must be added the will to forgive.

Ellul's Theory of The Three Milieus

In order to lay a historical foundation for examining Girard's theory, which is essentially metahistorical, we need to consider French historian and sociologist Jacques Ellul's theory of the three milieus. Ellul's concept of milieu is as precise as permitted by such a broad phenomenon.[12] A milieu is an environment, at once both material and symbolic, in which humans face their most formidable problems and from which they derive the means of survival and the meaning of life. A milieu has three basic characteristics: immediacy, sustenance and peril, and mediation. We are in immediate and direct relationship with our milieu; it forces us to adapt, to conform, just as surely as we manipulate it. From the milieu we derive all that we need to live—sustenance for the body and the spirit: food, clothing, shelter, order, and meaning. Concurrently, however, the milieu is the greatest threat to

9. Girard, *Scapegoat*, 79.
10. Girard, *Scapegoat*, 113.
11. Girard, *Things Hidden*.
12. Ellul, *What I Believe*, 99–103.

human existence, as in pestilence, famine, poisons, wild animals, political strife, war, pollution. The milieu, then, is *ambiguous* in value and produces an *ambivalent* reaction on our part—attraction and revulsion, desire and fear. This, I think, is the fundamental reason for the ambiguity of the sacred, which will be discussed in greater detail later.

A milieu is composed of two basic ingredients: meaning and power. Insofar as it is symbolic, a milieu is a human creation. The power of the milieu is harnessed for human ends; still, as an objective power it is not fully or even largely under conscious control. The dominant power of a milieu assumes one of three forms: nature, society, or technology. In the first instance, the power is a given. We exist in relation to the power of nature, and about this relationship symbolization first occurs. The power of society and that of technology (as a system) are human creations, but they are experienced in holistic fashion as objective forces; they are reified. In Ellul's theory, then, humans have lived in three milieus—nature, society, and technology. Yet this is no finalist theory of history in which the third stage represents the culmination of history; moreover, there is no deterministic principle that underlays the process.

Each subsequent milieu, e.g., society in relation to nature, mediates the preceding one, rendering it an indirect force. The preceding milieu becomes an ideological model for the subsequent milieu, thereby providing an illusion of where power resides. In dialectical fashion, however, it is actually the subsequent milieu that is used to interpret the preceding one. For example, in the milieu of society, nature is actually read through society; it is anthropomorphized.[13] Therefore, nature as a model for society is to a great extent a nature that is already a reflection of society.

To these three characteristics of a milieu we can add one that is implicit in Ellul's formulation: regeneration or renewal. Each milieu has a different principle by which it is regenerated. For nature the principle of regeneration is the feast; for society, it is sacrifice; for technology, it is consumption.

The chronology of the three milieus can only be approximated, of course. The milieu of nature or the prehistorical period is the hardest to date, in part because one runs up against the problem of when the human species (in the modern sense of the term) emerged. Archaeologist Colin Renfrew argues that there is some evidence, although still fragmentary, that the extensive use of human language to interpret the world metaphorically and to think conceptually is linked with the development of *Homo sapiens sapiens*, at least 40,000 years ago.[14] *Homo sapiens sapiens*

13. Kelsen, *Society and Nature.*

14. Renfrew, *Archaeology and Language*, 274–75.

may have emerged in the physiological sense even earlier, in some areas perhaps even as far back as 90,000 years ago. If one uses 40,000 years ago as a starting date for the milieu of nature, the efflorescence of the milieu of society (the historical period) can roughly be dated to 3,000 BC, the time of the rise of towns, civilizations, and literacy. The Neolithic period represents a transitional period from the late prehistoric period, the milieu of nature, to the historic period, the milieu of society. Finally, the post-historic period, the milieu of technology, can be dated to the post-World War II period and the widespread use of the computer. The transition to the milieu of technology includes the last two centuries.

The Milieu of Nature

The first thing to note about the milieu of nature is that of the three milieus we know the least about it, for here we are limited mainly to archaeological evidence. Myth and ritual, as they are known to us, come from societies already in the milieu of society or in transition to it. Especially in the earlier stages of the prehistoric period, humans were overwhelmed by a nature whose power appeared unlimited. Both technology and language were relatively underdeveloped. By contrast, the practice of magic was extensive. One might characterize the human relationship to nature as largely magical.

Human groups were small and widely dispersed. The low population density helps to explain the apparent absence of warfare between the groups. Archaeological evidence indicates that the bones of those of warrior age did not suffer wounds.[15] Hunting and gathering groups are marked by a relative absence of internal violence as well. Backed by neither an inequality nor a centralization of power, authority is specialized and personal, peculiar to one's talents.[16] Such authority, moreover, is easily contested or ignored.[17] Internal conflicts can sometimes be resolved by breaking up and regrouping.[18] Moreover, the redistribution of food in a situation of anticipated abundance mitigates some of the impact of internal strife.[19]

Ellul argues that it is plausible to extrapolate to the prehistorical period certain characteristics present in the historic period.[20] In this regard he applies Dumezil's theory of the three functions to the milieu of nature,

15. Ellul, *What I Believe*, 105.
16. Kottak, *Cultural Anthropology*, 103.
17. Turnbull, *Forest People*, 109–25.
18. Barnow, *Cultural Anthropology*, 256-67.
19. Sahlins, *Stone Age Economics*; Bird-David, "Beyond."
20. Ellul, *What I Believe*, 114.

maintaining that the earliest principle of organization for the group was not that of kinship but that of function. The three functions, understood more as principles of classification which together form a system, include religious and political sovereignty, military force, and material well-being.[21] Everything else of importance—the family, groups, taboos, cults, myth, and ritual—was a way of facilitating the performance of the function.

Cave paintings which date beyond 30,000 years ago indicate that the creative process of symbolization was well-established in this period. The subject of many of these paintings was the animal, which represented simultaneously danger and sustenance.

This is absolutely essential, I am convinced. The life milieu places one in a fully ambiguous situation: it is the source of life (physical, cultural, psychological) and the greatest threat to life. Cassirer's theory of the origin of language is consistent with this idea, as is Wheelwright's theory of metaphor (a type of tensive language) as a reflection of a reality experienced in terms of sharp contrasts, conflicts, and tensions that form a larger unity.[22] The ambiguity of sacred value as expressed in myth and ritual is a consequence of the ambiguity of life milieu.

Finally, there is no sense of good and evil in the milieu of nature. Moral evil was experienced in the same way physical misfortunes such as disease, famine, suffering, death were—as "moments of the cosmic totality."[23] Good and evil, as we shall see later, are categories peculiar to the milieu of society.

The Milieu of Society

The milieu of society or the historical period dates to the emergence of towns around 3,000 BC, although the Neolithic period represents a transition from the milieu of nature to that of society. He identifies five characteristics of this milieu.

The first is an increasing reliance on artificial means of action that allow humans a certain freedom from the determinations of nature. Voluntary action becomes self-reinforcing. Among those phenomena most responsible for the growth of the artificial and the voluntary are technology, literacy, law, and the entire range of social institutions. Technology allows humans both to distance themselves from nature and to master it. Literacy improves one's ability to think analytically, for now the ideas can be detached from their

21. Littleton, *Comparative Mythology*, 3–6.

22. Cassirer, *Language and Myth*; Wheelwright, *Metaphor and Reality*.

23. Eliade, *Quest*.

source, reread, and reflected upon. Writing permits a thought or decision to be applied to a variety of contexts. The absoluteness of cultural forms that functioned as archetypes can now be questioned.

Law was undoubtedly a central force in creating a society increasingly distinct from nature. A formal system of laws that begins by running parallel to the extant taboos and imperatives becomes increasingly important and at times is in conflict with religious norms. The creation and manipulation of law resulted in the main concern in the milieu of society: politics.

Not just law, however, but all the social institutions as they become more rationalized and complex work to produce, at least among the elite, a sense of the artificiality of society. Politics is an implicit recognition of this.

The second characteristic is the influence of one society on another. The emergence of towns facilitated this, as did literacy and the proliferation of and interest in artifacts. In the prehistorical period, by contrast, the human groups were too dispersed and their oral and visual symbolism too group-specific to afford transference to a different context. It is, of course, both the strength and the limitation of literacy that it gives rise to a higher degree of abstraction and objectification.[24]

Social hierarchy is the third characteristic. Following Dumont, hierarchy is almost the opposite of social stratification in the modern era.[25] Hierarchy assumes a view of society as an ordered whole so that each division within society, whether based on gender, age, class, occupation, or moiety, has a *complementary* relationship to its opposite. Moreover, a hierarchy, at least ideologically, is concerned more with status than with power.

Ellul speculates that the aristocratic class's claim to have an eponymous ancestor was the source of its authority. The eponymous ancestor provided the society with both a history and a hierarchy, for no social category exists alone but only in relation to its complement. The ancestor provides not only a history of descent but also a history of events, and, finally, a theory of authority within society. As Ellul puts it: "The aristocratic families that preserved the memory of their history also had another essential role. As a frame for the whole people, they carried in effect the history of the entire society. Their history was the history of all. They were truly the memory of the society."[26] This confirms Dumont's point that in a hierarchy the term higher in status is more representative of the whole than its lower status complement. Hence the aristocratic families are more representative of the society than the non-aristocratic ones.

24. Ong, *Orality and Literacy*.
25. Dumont, *Homo Hierarchicus*.
26. Ellul, *What I Believe*, 119.

The fourth characteristic of the historic period is the institution of law (previously mentioned under artificiality). Law as a general organizing principle is a consequence of the artificial and voluntary being brought to bear upon problems created by social hierarchy, problems that myth and ritual did not cover. Ellul argues that law was a response to three challenges: space, time, and human relationships.

Law allows humans to tame symbolically a section of the natural environment, to establish boundaries between natural space and social space. Law permits society to become the human life milieu. Law likewise responds to the challenge of time, flux. As Eliade has demonstrated repeatedly, archaic man did not want to make history but rather to return to the time of creation. And yet the more humans introduced technology, literacy, and rationality into their relationship to nature, the more unpredictable life became. There were unintended social consequences to every artificial and voluntary attempt to control nature. Law established the regularity in social relationships that myth and ritual were by themselves unable to provide anymore.

In the historic period the increase in the size of societies poses a threat to human relationships because one cannot personally know everyone else. Moreover, as politics is slowly disengaged from religion, the exercise of power is problematic. Law acts to regulate relationships between individuals and between kin groups.

Creativity and diversity in meeting the challenges of life is in particular a characteristic of the historical period. Not that every response is adequate, as Toynbee has shown. The diversity of responses in meeting the challenges of nature and society precludes any principle underlying history. Among the creative responses that are most significant are those of compensation. Ellul suggests that every leading phenomenon, whether it be an institution, class, or group, that is instrumental in both regulating society and helping to meet natural and social challenges wants to become all-powerful. This movement toward totalitarianism is accompanied by compensations of two major types: (1) increase in holidays, sports, and play; (2) religious escape, as in mysticism, festivals, and sects. If the society in question is experiencing disorder, even anarchy, the compensations will often prove to be the emergence of a strong, centralized government or a rigidly organized and heavily legalistic religion.[27]

The most pressing and profound problems in the milieu of society are political—relations between groups, internal administration, and warfare. Natural problems such as wild animals, poisons, disease, and famine are

27. Ellul, *What I Believe*, 126.

still present and for a period can become the most dangerous; in the long run, however, the organization of society and relations between societies are the greatest concern.

Ellul maintains that each subsequent milieu mediates the preceding one rather than simply eliminating it. Therefore in the historical period, society mediates nature. This is most forcefully demonstrated in Hans Kelsen's masterful *Society and Nature*. Nature is interpreted as a society, a more powerful one, to be sure. I shall use Kelsen's main conclusion on this issue as an illustration. In the milieu of society or transition to it, the principle of retribution was seen to underlay all of nature: every good deed is rewarded, every evil deed punished. Obviously the principle of retribution is the first theory of legal punishment and as such is projected onto nature, a bigger and better society.[28]

At the same time nature as a more perfect society becomes an ideological model, a model invoked after the fact to justify decisions and actions that are actually social (political). A good example is the concept of natural law, which emerges in the historical period.

The Milieu of Technology

Occurring in the posthistoric period, the technological is the third type of milieu. Strange as it may seem, modern technology fits Ellul's definition of a milieu: that which is concurrently the means of life and the greatest threat to life, that which is most immediate to us, and that which mediates all our relationships. The first point is abundantly clear today; we look to technology to solve all our problems. One interpretation suggests that the concept of social problems is the dominant modern symbol of evil.[29] To regard evil metaphorically as a problem is to adopt the technological outlook: all problems can be engineered away. Yet in warfare, pollution of the environment, genetic and chemical manipulations, propaganda and the mass media, and bureaucracy, technology poses an enormous threat to our physical and psychological well-being.

Technology is much more immediate to us than nature, for in modern societies our physical environment is one of plastic, glass, steel, and concrete taking the form of an enormous diversity of technological objects and consumer goods.

The other side of this is that technology mediates all our relationships. Whether for purposes of production or consumption, technology

28. Kelsen, *Society and Nature*.
29. Stivers, *Evil in Modern Myth*, 48–75.

mediates our relationship to nature. By both material and non-material technologies we reduce nature to one enormous amusement park. Even more importantly technology mediates our relationships with each other. Human techniques (techniques for the manipulation of human beings) tend to supplant morality, manners, and social institutions. Bureaucracy (a technique of organization), propaganda, advertising, public relations, psychological techniques (such as the "how to" books on childrearing or becoming successful or having a positive outlook), and expert systems are only some of the most obvious ones.

Such techniques in the service of efficiency and power reduce the human being to an abstraction, the object of a method that makes everyone the same. But technique has pernicious consequences for the user as well as the recipient of technique. As an objectified, rationalized method, technique does not depend upon one's subjectivity, one's experiences. As Illich has shown, all of life is given over to experts. The wisdom of age and experience is given scant attention over against technical expertise.[30]

Ellul's objection to technology is not philosophical; it is historical. Until the nineteenth century technology developed at a pace that allowed it to be integrated into culture; technology was subordinate to culture. Because of an unbounded faith in technology and a conscious intention to experiment with technology and find a diversity of uses for a single technique, technology came to dominate culture.[31] Inappropriate uses are a consequence of the diffusion of technique into every sphere of life: all the efforts to measure and rationalize that which is qualitative (dependent upon context).

As technology proliferates, has generalized applications, and becomes coordinated (through the computer), it tends to form a system. Techniques are related to one another before they are adjusted to human use and need. Humans invent technology, but insofar as it has become a kind of system (without true feedback), they allow it to function autonomously. Technology as a system (a system of techniques coordinated by computerized information) is superimposed upon human society, which at least in the milieu of society was organized by the symbolically mediated experiences embedded in cultural institutions. Technology foments a crisis for culture because it renders it ephemeral and transitory.

Technology's impact upon the milieu of society can be ascertained by examining what has happened to social hierarchy, the law, and the creation of symbols. Social hierarchy (in Dumont's sense of the term) has

30. Illich, *Disabling Professions*, 1–39.
31. Ellul, *Technological Society*.

disappeared in modern societies.[32] Replacing it is that to which the concept of social stratification refers—a culturally unintegrated set of classes whose basis of domination is power and not status. Dumont argues that there is no general ideology that supports inequality in modern societies; indeed, the modern ideology supports the values of equality and individualism. Therefore, a set of classes exists with, as it were, a bad conscience, and without bonds of reciprocity and complementarity.[33]

It would appear that the concept of mass society provides the explanation for the existence of competitive classes without full cultural justification. Tocqueville and Kierkegaard understood this phenomenon well before social scientists of the 1940s and 1950s made use of the idea. A mass society is one which is paradoxically collectivistic and individualistic at the same time. Cultural authority, which in historical societies resided in hierarchy, declines, along with kinship and the community as institutions that implement it. Thereupon authority, which has become power, resides in the state and common opinion. The individual, isolated from and fearful of his fellow citizens, to whom he can no longer turn for aid, because in an age of equality obligatory norms of reciprocity wither away, turns instead to government for help. The power of the state and public opinion grow stronger and stronger with the collapse of hierarchy to meet the needs, demands, and fears of the isolated individual. At the same time, however, the individual believes that she can become an individual through the efforts of the state.

The growth of state power and public opinion has been made possible through technology so that the modern political state is largely technical, i.e., bureaucratic. Likewise, the formation and measurement of public opinion has become almost exclusively a technical phenomenon. Ellul has argued that politics itself has largely become a matter of rubber stamping decisions actually made by bureaucratic and technical experts in the interest of efficiency.[34] Politicians' choices are always about what is technologically *possible*. On those occasions when politicians appear to resist technology, it is only by stopping technology from doing something positive, e.g., stopping the growth of nuclear power plants for civilian energy usage. But these occasional negative decisions are much less important than the positive commitment to technological growth. Yet traditional politics most often involved compromise, which is the antithesis of a solution.

It would be absurd to claim that law has disappeared in the milieu of technology. If anything, it is proliferating at a geometric pace. But as numerous

32. Dumont, *Homo Hierarchicus*, 256–58.

33. Dumont, *Mandeville to Marx*.

34. Ellul, *Political Illusion*.

studies have indicated, this development has little to do with the value of justice.[35] All forms of law and not just administrative law have become largely technical; their principle concern is efficiency. Bureaucratic rules are, strictly speaking, amoral, for they are technical in nature.

The most problematic aspect of the milieu of technology is the erosion of the symbol as an active force in human evolution. This is a result of two interrelated phenomena: the technological system and the universe of material visual images (especially in the mass media).[36]

The visual images of the mass media have had a corrosive effect upon natural language. It is not due mainly to the speed and sheer quantity with which the images hit us. Rather it is a result of the formation of a pseudo-language accomplished through the logical sequence of visual images in a program unit and the subordination of the symbol to the material visual image. Debord, Baudrillard, and Ellul have shown that the media present to us a more emotionally satisfying reality than the reality of our lives.

The visual images of the media decontextualize reality by removing it from its existential, cultural, and historical contexts. (A symbol only has meaning in context.) The visual images impose upon us a strictly material and "objective" reality, but one that has been *selectively* represented. The visual images deconstruct and then reconstruct reality according to a spatial logic of signifiers: images of power and of possessions. We see objects to be possessed and consumed, the power of which becomes that of the consumer. These acts of power are spatially linked together in a television program or movie in a logical sequence that leads to an outcome—success or failure. "Meaning" is reduced to power.[37]

Visual images become the operational indicators of words. Advertising is adroit at reducing aesthetic and ethical qualities to consumer goods and services. Love is holding hands and drinking a Coke. Once again, a strictly material reality.

The use of symbols in advertising and political propaganda leads to a subjectivization of the symbol. Verbal symbols are ephemeral today; they no longer motivate us nor have utility in a technological civilization. For purposes of advertising and propaganda, they have largely been rendered meaningless. If all natural languages are polysemic, the context (both as sense and reference) allows for a certain precision of meaning. In advertising and propaganda, by contrast, the polysemy of a symbol is so stretched (the tension between sense and reference breaks) that it can be applied to

35. Blumberg, *Criminal Justice.*
36. Ellul, *Humiliation of the Word.*
37. Stivers, "Deconstruction of the University," 120.

virtually anything. For instance, the term *democracy* is used by politicians to refer to every extant government (if it is on their side). As Kenneth Hudson demonstrated, when discourse is not used for purposes of factual description and technical operationalization, it is increasingly used for purposes of emotional identification.[38] Advertising and propaganda use symbols to create an emotional bond between the product and its users.

This is an enormous paradox. Natural language, which is being stripped of its traditional meaning and thus made subjective (in an emotional sense), is concomitantly made objective through its association with the subordination to material visual images. The fundamental reason for this incredible transformation is that symbols have lost their practical purpose in a technological civilization.

In the historical period, in the milieu of society, symbols served two related purposes. First, important historical events became symbolic of some action, characteristic, or quality a society wished to retain or to avoid by preserving the memory of its occurrence. Symbols gave definition to and provided a cultural identity for a group. Second, symbols were the primary ingredients of social institutions along with the experiences they made sense of. Yet technology and the material visual images that represent it are now supplanting the symbolically mediated experiences embodied in social institutions.

Ellul's theory of the three milieus in no way suggests that the technological milieu is the final milieu or that we cannot return to the milieu of society. It does, however, point to the major problem in the technological milieu—the eventual loss of freedom and creativity. For it is only within a vigorously symbolic language that humans imagine real alternatives and exercise moral freedom. Without this, the ability to make history declines: time becomes the repetitive time of technological progress.

Each milieu depends, of course, upon the reciprocal, socializing actions of humans to carry out the tasks of survival and the search for meaning. Humans are markedly different, however, in each milieu. Humans become distinctly human (with a sense of history and of the freedom to create a future) only when society and nature are viewed in dualistic terms and when society and technology are similarly perceived; this occurs only in the milieu of society or the transition to it. In the milieu of nature there is no sharp distinction in kind between human and animal; both are part of nature. Certain animals that are more powerful or more central to a group's survival become the emblems of the milieu (totem animals, for instance). In the milieu of technology as well there is in practice no

38. Hudson, *Jargon of the Professions*; Hudson, *Language of the Teenage Revolution*.

significant difference between the human and the technological object. Here too certain powerful objects, such as the automobile and the computer, become representative of the milieu.

Humans in the milieu of nature and milieu of technology not only occupy a position of lesser importance but do not experience (ideology to the contrary) themselves as separate and autonomous agents. Only through a sense of history and of the freedom to create a future do humans achieve a relative autonomy in regard to nature and technology. As socializing agents of their culture, individuals in the milieu of nature and the milieu of technology indirectly act as representatives of powerful forces in nature and technology respectively. In this regard, they are not, strictly speaking, *human* exemplars.

Anticipating later discussion, I maintain that only in the milieu of society is mimetic desire directed toward human models. In the milieu of nature, mimetic desire is directed toward natural objects, e.g., animals; in the milieu of technology, toward technical objects (which may include humans). All desire is mimetic desire: each milieu because of its enormous power and importance to our existence and because of our symbolic response to it directs desire toward itself.

The Feast, The Hunt, And The Scapegoat

The following remarks, adventuresome as they are, suggest that the theory of the three milieus may help to resolve certain issues in the study of comparative religion. Foremost among these, for our purposes, is the controversy about whether hunting or scapegoating is the origin of ritual. As a prelude to this four other conundrums are briefly discussed: nature and society as represented in myth, the general meaning of witchcraft, the disappearance of the high god, and changes in the content of cave art. Later the theory of the three milieus will allow us to place Girard's theory of mimetic desire and ritual scapegoating into historical context and will yield a theory of the sacred festival.

Jonathan Z. Smith[39] has criticized Eliade[40] for not appreciating the extent to which the mythical treatment of sacred space and sacred time has society as its real subject. Smith reinterprets an Australian cosmogonic myth belonging to the Tjilpa tribe of the Arunta.[41] Eliade had placed the narrative of the Tjilpa into the theoretical context of *The Myth of the Eternal Return*.

39. Smith, "Wobbling Pivot"; Smith, *To Take Place*.
40. Eliade, *Sacred and the Profane*.
41. Smith, *To Take Place*.

Archaic man, according to Eliade, did not wish to make history, to change, but rather to live in the time of perfection, the primordial totality of the chaos that precedes creation, or to live in the time immediately after creation, the time of the ancestors. Whichever primordiality is given greater emphasis—chaos or creation—there is the understanding that creation had to take place.[42] The aboriginal events of creation become archetypes. In Eliade's famous formulation, archaic man wants to live as close as possible to sacred space, the center, the place where the creation of the cosmos occurred.[43] The center, then, was symbolic in all its forms—the cosmic mountain, the temple, the center of the village, the center of the house, the pole. In the Tjilpa myth the broken pole is thought to symbolize the end of the world, which, because it precedes recreation, also occurs at the center.

Sacred time is the time of creation: the movement from chaos to creation. For Eliade the narrative of the origin of one's ancestors is *subordinate* to the myth about the creation of the cosmos; society is read through nature.

By contrast, Smith understands the myth correctly, I think, to be a narrative about the society of ancestors and how they transformed the landscape. It is a symbolic *history* in which the events, recounted as a chronology of travels, become symbols by leaving traces of the ancestral wanderings across the physical environment. These traces are the remains of the ancestors themselves metaphorically present in objects and tools. Hence the ancestors' remains populate the environment. The meaning of the broken pole in Smith's reinterpretation is *social*: the ancestors' broken pole is symbolized today in a tall stone. Smith agrees with Geza Roheim that in the various narratives there is the sense that the environment is a human *construction*.[44]

Smith likewise takes on Eliade's interpretation of the temple in ancient Near Eastern mythology. Eliade suggests that the temple is metaphorically assimilated to the sacred mountain, the center of the world, the place of creation. Apparently there is no sacred mountain, at least as far as the present evidence goes. Smith proceeds to demonstrate that the temple has a primarily political meaning as the locus for the performance of the king's duties. Finally, Smith demonstrates in relation to the same cosmogonic narratives that the act of creation is an essentially human one—the act of *building a physical structure*.[45]

Smith implicitly recognizes the milieu of society in his reinterpretation of sacred space as social space represented in mythology. As I mentioned

42. Eliade, *Quest*, 86–87.
43. Eliade, *Myth of the Eternal Return*.
44. Smith, *To Take Place*, 6–9.
45. Smith, *To Take Place*, 13–21.

earlier, the only surviving myths are from the historical period, when society had become the principal human milieu.

Hans Kelsen's *Society and Nature* shows how every object in nature has a *social* relationship to other objects. For example, in countless myths around the world the relationship between planets is marital (intercourse), familial, cooperative (between friends), or competitive (between rivals), and they express a full range of human emotions. The dualistic conception of society and nature is actually a rather late development in the historical period. Early in the period, at least in the West, nature and human society were both part of a universal society ruled by the principle of retribution; there was as yet no dualism. Kelsen notes a sharp distinction, however, between archaic Greek theology and medieval Christian theology. For the former the principle of retribution meant that divine law could be violated but each violation was met with sure punishment. Here natural law is penal law. For the latter nature is ruled by a divine law that cannot be violated (the law of causality that derives from the principle of retribution is at work); because of the conception of human freedom and the perceived need to develop a theodicy, society becomes the arena in which divine law can be violated. The dualism of nature and society is now complete.[46]

Rene Girard has interpreted witchcraft, belief and practice, as a form of ritual scapegoating. Certainly the practice of witchcraft proliferates and intensifies to the extent that internal strife or external conflict grows strong.[47] Moreover, the witch represents an even more striking anthromorphism than the pantheon of deities does. For both of these reasons witchcraft belongs in the milieu of society. Colin Turnbull's *The Forest People* shows that the Mbuti Pygmies have no belief in witchcraft or sorcery, whereas the villagers nearby have a great fear of it.[48] In my reading, the Pygmies exist in the milieu of nature (at least when allowed to), the villagers in the milieu of society. Witchcraft, then, is a major form that magic assumes in the milieu of society. Magic in the milieu of nature is not concerned with political and military problems; rather it is an expression of the attempt to imitate and renew nature. The Mbuti Pygmies provide an apt example in their lavish imitation of animals both before hunting and during the molimo festival.

The disappearance of the high god, the creator god, can readily be explained in terms of the subordination of nature to society in the historical period. Eliade notes that in some mythologies the creator god eventually

46. Kelsen, *Society and Nature.*

47. Douglas, *Natural Symbols*; Wallace, *Death and Rebirth.*

48. Turnbull, *Forest People,* 228.

become a *deus otiosus*.[49] He plays less and less an efficacious role in the daily lives of the religious society, perhaps only remembered for the work of creation and called upon solely in times of crisis. As Eliade perceptively observes, the transformation of a high god into a *deus otiosus* occurs in groups for whom sacred history takes precedence over cosmic time (the passage from chaos to the creation of the cosmos). The former kind of primordiality is directly linked to the milieu of society. The high god is relegated to the attic of collective memory, as lower deities (often ancestors) who play an active part in the sustenance of the group supplant him. Society and its history supersede nature and its cyclical time although the two are intertwined in the same myth. Nature is relegated to *ultimate* origin while the ancestors and their history become the *immediate* origin.

Striking empirical evidence for the idea of a transition from the milieu of nature to that of society is present in cave art. A French archeologist's analysis of cave art indicates that 95 percent of the Paleolithic representations were of animals; furthermore, the few human figures were without arms. By the Neolithic period (the transition to the milieu of society) depictions of human groups, accompanied by tools and weapons, became pervasive.[50] The animal is a dominant figure in the Paleolithic period as threat to life and as sustenance (both in real and symbolic ways); by the Neolithic period the animal is under human domination as the hunted. The focus of human representation has become the human group and its technology.

The meaning of the hunt is hotly debated today. Walter Burkert has made the ritualized hunt the cornerstone of his theory of religion; ritual killing in the hunt precedes animal sacrifice.[51] This is the inverse of Girard's thesis that sacrifice is the originary sacred act. Burkert accepts the hypothesis that hunting is the origin of human social organization in the Paleolithic era.[52] Hunting required cooperation, coordination of activity, and a division of labor. The emotional shock of the kill—the passage from life to death—as well as the realization that the aggression of the hunt was somehow related to male competition for females is superimposed upon an act essential for survival. Consequently the hunt is ritualized, and the energy and anxiety of the hunt can be sublimated.

Burkert concludes that there may have been two originary acts: the killing of the animal for food and the selection of one of the group to be an "offering" to predatory animals so that the rest of the group can escape. The

49. Eliade, *Quest*, 81–82.
50. Ellul, *What I Believe*, 106.
51. Burkert, *Homo Necans*; Burkert, "Problem of Ritual Killing."
52. Mack, "Religion and Ritual," 29.

former gives rise to the feast sacrifice, the latter to the aversion sacrifice. Only the latter can accurately be termed scapegoating. This is an enormously important distinction. [53]

The feast sacrifice, Burkert notes, involves ritual killing followed by a communal meal. He emphasizes the distributing and sharing of the meat as a universal phenomenon. Ritual killing and ritual eating become the religious representation of the acts of sharing and ingestion of sacred food—meat.[54]

The aversion sacrifice, a form of scapegoating, is the offering of another human or animal to persecutors to protect the group. Burkert claims that anxiety rather than mimetic desire is the psychological process behind the sacrifice.[55] In Girard's theory, of course, the persecutors are the other members of the group; here there are no external predators. By contrast, Burkert views this kind of sacrifice as a means to avert danger. The feast sacrifice is an "offering" to the good gods; the aversion sacrifice, an "offering" to the evil gods. Differences notwithstanding, the two forms of sacrifice share in common the act of ritual killing. Burkert observes that as yet he has not discovered a common strand of meaning that links together the two forms of sacrifice.

Burkert's critics point to the dearth of evidence for a ritualized hunt in the Paleolithic period; it would appear to be a product of agricultural societies.[56] Moreover, Adolf Jensen argues that ritual killing often involved domesticated animals. His main theory has two parts. First, prior to ritual killing, the animal was protected by taboos and rules that governed when and how animals were to be killed. The "master of the animals" protected animals by limiting the number that might be taken, leading the game to the hunter, and insuring their spirits could be reincarnated so that "there will always be fresh game." Mythology that accompanies such rituals often involved a series of disclaimers in which the hunters disavow responsibility for killing the animals; e.g., they maintain that the arrow went off course, or claim it was for the animals' own good.[57] Second, ritual killing or sacrifice that appears later with the rise of agricultural societies is a celebration of the act of killing and represents it as an act for renewing life. This renewal of life is based on a primal event, the murder of a deity who subsequently became the source of

53. Burkert, "Problem of Ritual Killing," 173–74.

54. Burkert, "Problem of Ritual Killing," 173–78.

55. Burkert, "Problem of Ritual Killing," 178.

56. Mack, "Religion and Ritual," 29.

57. Jensen, *Myth and Cult*, 138.

fecundity. As Jensen puts it, the core idea of ritual killing is that "all animal life . . . must maintain itself by the destruction of life."[58]

Both Burkert's distinction between feast sacrifice and aversion sacrifice and Jensen's between taboos protecting animals and ritual killing are especially helpful when reinterpreted according to the differences between the milieu of nature and the milieu of society. I suggest that there are three distinct but not mutually exclusive kinds of ritual that have as their subject either food or sacrifice; moreover, I think that, considered as a sequence, they mark the passage from the prehistoric to the historic period.

The first is the feast, the communal meal of religious significance. I maintain that the *food* was sacred, and the animal in particular. The *killing* became ritualized at a later point. In the milieu of nature, the human group was relatively weak in relation to nature and the level of technology primitive. With respect to wild animals there were two dangers: physical harm and their disappearance as food. A milieu as well has a principle of regeneration. Jensen claims that the central assumption of ritual killing is that the taking of another life is necessary to sustain one's own life. Could we not conjecture that even more primitive than this is the existential experience that from death comes life—the principle of regeneration in the milieu of nature? If this is valid, then given the human position of subordination to nature, the ritualization of killing would wait until humans (perhaps in the Neolithic period) began to exercise greater dominion over their physical environment. The food, the animal, is more important than the act of killing, especially if one did not actually perform it. The eating of the animal is a kind of *sacrifice*, but in the sense of *renewal*—the renewal of human life and of nature itself.

In *The Forest People*, Colin Turnbull demonstrates that, for the Mbuti Pygmies, who were hunter-gatherers, there was no ritualized killing. The molimo, "the animal of the forest," was sacred and the center of their great festival. The Mbuti offered the molimo food, but concurrently they feasted for a month or more on the same food (that was actually the molimo). The eating of the molimo is the heart of the festival: "the real work of the molimo . . . is to eat and to sing, to eat and to sing."[59] Turnbull's ethnography offers support not only for the argument about the feast but also for the entire thesis about the milieu of nature.

Maringer's insights about primitive funerary feasts, conceivably "memorial or sacrificial meals," are apt. These and other rites he surmises sought

58. Jensen, *Myth and Cult*, 190.
59. Turnbull, *Forest People*, 80.

protection from the dead and assistance in the "quest for food."[60] Leroi-Gourhan agrees on this point and has suggested that Paleolithic cave art involves a complex system of symbols that center on the idea of fecundity.[61] The feast and the funeral—how appropriate that they be associated according to the principle of regeneration. The words feast and festival come from the same root; the feast is the first festival.

The second and third rituals involve ritual killing: the ritual killing of animals (mainly domesticated) on the one hand, and ritual scapegoating (involving the ritual killing of humans) on the other. My argument, in brief, is that the mythological justification for the ritual killing of animals comes from the ritual scapegoating of humans. Moreover, both kinds of ritual killing (assuming with Girard that ritual scapegoating originated in an act of murder) are a much later development than the feast and, perhaps, occurred simultaneously. Insofar as hunting was ritualized without killing it was part of the feast sacrifice, but when it was assimilated to ritual killing it took on a different meaning (aversion sacrifice).

The ritual killing of animals and the ritual killing of humans begin in the transition to the milieu of society. Both hunting and the ritual killing of animals take on a deep symbolic meaning. First, they symbolize human mastery of the physical environment, just as the later cave art does. The less dependent humans are upon hunting as a source of food, the more symbolic value it has; inversely, the more dependent humans are upon hunting the greater the symbolic value of the food itself (as in the milieu of nature). Second, and more importantly, the ritual killing of animals is also assimilated to the principle of regeneration in the milieu of society—human sacrifice or scapegoating.

Jensen argues that there is a continuity to ritual killing, but that the meaning must be sought in the oldest available myth.[62] Only with the original meaning can we unlock its mystery. Of interest here is not his theory of semantic depletion, by which a ritual loses its original significance but still has useful application, but the view that ritual killing is originally a recollection of the slaying of a demi-deity from whose body spring most forms of life—human, animal, and root crop. The slain deity is, in Girard's theory, the ritual scapegoat who is later mythologized as a divinity. If Jensen is correct about the unity of ritual killing, then the ritual killing of animals makes *full* sense only when it becomes a variation of or substitution for human sacrifice.

60. Maringer, *Gods of Prehistoric Man*, 29.

61. Cited in Ucko and Rosenfeld, *Paleolithic Cave Art*, 143.

62. Jensen, *Myth and Cult*, 166–68.

What I am getting at in this apparent digression is that Girard's theory of ritual scapegoating is correct for the milieu of society. Only in the milieu of society do politics and morality become a dominant concern. If the milieu of nature is organized around the issue of life and death, the milieu of society is structured by the theme of good and evil. Eliade was right on the edge of this discovery when he contrasted the cosmic polarities, such as life and death, with human polarities, such as good and evil.[63]

The original meaning of the feast and the hunt in the milieu of nature is now overlaid with the meaning of sacrifice (in Girard's sense). This occurs because society mediates nature but does not reduce its stature. Nature remains a powerful force in human life, so much so that nature and society become symbolically an almost impenetrable ambiguity. Society (in the sense of hierarchy, retribution, etc.) is projected onto nature, but nature remains an ideological model (as with the concept of natural law). The sacred history of the ancestors becomes more important than the creation of the cosmos, but the form of time (except for Jews and Christians) remains the cyclical time of nature.

The Festival: The Sacred, Regeneration, And Desire

In the previous section I unequivocally accept Girard's thesis of ritual scapegoating as the origin of society. Furthermore, I agree with his theory of mimetic desire and the nexus between the two. I do not, however, think that the theory of mimetic desire and ritual scapegoating applies either to the milieu of nature or to that of technology. This in no way takes away from Girard's achievement; rather it establishes its temporal boundaries. I have already made a step in this direction by arguing that the feast is the principle of regeneration in the milieu of nature, just as sacrifice is in the milieu of society. Let us now examine the sacred, desire, and the principle of regeneration in each of the three milieus as a way of developing a theory of the festival.

The sacred is in the most general sense the life-milieu. What better fits Eliade's definition of the sacred as power and reality than the milieu? That the milieu places us in an unresolvable situation of nurture and danger is the source of the ambiguity of the sacred. Caillois's theory of the sacred explores this basic ambiguity: the sacred is both that which is holy and unsullied and that which is evil and defiling; likewise the profane is that which is evil and defiling and the neutral. His deft linguistic analysis reveals that rather than two terms—sacred and profane—there are actually three terms—sacred of

63. Eliade, *Quest*, 174.

respect, sacred of transgression, and profane. The term sacred of transgression is *implicit* in the natural languages he examined; hence it existed without conceptualization. The idea of a sacred of transgression is embedded in the meaning common to the ambiguous terms *sacred* and *profane-defilement*.[64] The fundamental tension a milieu creates is responsible for the ambivalence toward the sacred: attraction and repulsion, fascination and fear.[65]

Both Caillois and Eliade understood the ambiguity of the sacred as being both chaos and creation, with the latter emerging from the former and requiring periodic renewal. Nature was their model of the sacred because they assumed that it was the model of the people they studied. Moreover, they understood that nature and secondarily society consist of two opposing forces so linked that the renewal of the milieu involved the movement from the one pole (chaos, evil, death) to the other pole (creation, good, life). What is revealed here is the experience of a milieu as a dynamic rather than a static phenomenon. The milieu exists and renews itself by virtue of the passage from one pole to its opposite.

In the prehistoric period, nature is experienced as sacred, manifested in specific hierophanies. The two cosmic poles that organize the milieu of nature are life and death. The principle of regeneration is the feast, the ritualized meal. Food (especially meat when available) and eating symbolize the movement from death to life and permit a certain communion with what is eaten. The more deeply immersed humans are in the milieu of nature, the more important the ritual meal is over against the act of killing. The festival, *which is the ritualization of the experience of regeneration*, is, in this instance, the feast.

In the historic period, both nature and society are experienced as sacred, but nature has the face of society. Nature is read through society but concomitantly nature is the ideological arena onto which social conflicts are projected. The deities in nature face the same moral and political tensions and conflicts that humans do. Sacred history, the history of the ancestors, becomes more important than the cosmogony. Eventually the earthly king becomes sacred and may even be seen to be the ruler of nature. The aristocratic class may be viewed as quasi-sacred, as descendants of the ancestors. Finally, the church as a religious society may unwittingly be regarded as sacred.

Politics and morality arise in the milieu of society to the extent that the most pressing problems become moral. In this milieu the polar tension is between good and evil. Strictly speaking the terms sacred and profane only

64. Caillois, *Man and the Sacred*.
65. Otto, *Idea of the Holy*.

arise in this milieu and have a decidedly ethical cast. Eliade observes that, until the undesired parts of life were regarded as *evil*, they were "accepted as constitutive and unexceptionable moments of the cosmic totality."[66] The idea of retribution is seen to be universal, encompassing both nature and society. Natural disasters and failures become a punishment for transgressions. Girard's theory accounts for the principle of regeneration in the milieu of society—human sacrifice. Sacrifice as scapegoating represents the expulsion of evil. Caillois implicitly understood this when he said that a theory of the festival would have to be "correlated with a theory of sacrifice." Sacrifice, he observed, "is akin to the inner mechanism that sums it up and gives it meaning."[67]

The festival is enlarged now to include, in addition to the feast and the hunt, scapegoating and every form of transgression—the violation of taboo, role-reversal, and so forth. To chaos, suffering, and death is now added moral chaos or evil. We may even speculate that the taboos surrounding the hunted as embodied in the "master of the animals" are the imposition of moral norms within the milieu of society on a much earlier and essentially amoral activity. The taboo is itself an artifact of the milieu of society.

Insofar as food, animals, and survival remain a central and immediate concern, society has to be accounted for mythologically from nature, e.g., out of the body of the slain demi-deity, and the form time assumes continues to be that of nature, even though the content is the sacred history of the tribe. Nature is reduced to the distant origin of society and, like the high god, its paramount symbol, it recedes into the background as the more immediate and serious political and moral issues occupy the foreground.

In the posthistoric period both society and technology are experienced as sacred, but now society is read through technology. The paramount political and moral problems are being transformed into technological problems to be solved by technical experts, e.g., the medicalization of moral issues.[68] Technology is both power and reality today. Although at one level technology (as rational, efficient, objectified method) is abstract, at another level it is concrete—technological objects, consumer goods. These act as hierophanies.

A technological society is one whose chief value, purpose, or goal is efficiency, maximum production and maximum consumption. Over against rational technique stands inefficiency as instinct, the will to power, eros. These opposite poles of a technological milieu are, however, at a deeper level related.

66. Eliade, *Quest*, 174.
67. Caillois, *Man and Sacred*, 185.
68. Ellul, *Political Illusion*.

First, as Ellul notes, technology and instinctual desire form a dialectic: desire today can only be satisfied by technology, and technology can only advance by the constant stimulation of appetite.[69] Jean Brun first called attention to the paradox that the cold, impersonal, abstract force of technology does not finally appeal to reason and moderation but to our desire for power and possessions. Technology as a system is the "head of Apollo" superimposed upon instinct, the will to power, the "body of Dionysius."[70]

The more reason becomes objectified and collectivized in the technological system, the smaller the role for subjective reason based on symbolically meaningful experiences becomes. This tends to enlarge the play of irrational or instinctual forces. Moreover, the cumulative impact of technology acts as a repressive force; the more technology demands of us in terms of regulations, schedules, and coordination, the more we apparently need to escape this kind of rationality by plunging into the irrational, into random sensations.[71] The result is a society that is at once both extraordinarily rational and irrational.

The instincts most associated with the will to power are sexuality and aggression. Indeed sex and violence can be interpreted as an ersatz sacred of transgression over against the technological system.[72] As forms of inefficiency they are the negative to the positive pole of efficiency (technology). If a milieu is comprised of two poles in tension, and the principle of regeneration involves the movement from the negative to the positive, then here that principle is *experimental consumption*.

The sacred power of technology becomes manifest in technological objects (consumer goods). These hierophanies of consumption become differentially sacred depending upon individual circumstances. We have already seen, however, that technology, while manifestly opposed to instinct, is perfectly suited to it at a deeper level because both represent the will to power. Moreover, advertising uses sex and violence to sell these consumer goods, for instance, the eroticizing of the automobile. Even more important than advertising's direct use of sex and violence is its indirect use: the consumer goods of advertising are placed in spatial relationship to the sex- and violence-saturated programs of the mass media. In this sense programs are an ad for advertisements. In consuming the technological object we are indirectly consuming the instinctual power of sex and violence. Enlarge the sphere of the instinctual, and the desire to possess and use technological objects increases.

69. Ellul, *Technological System*, 316–18.
70. Cited in Ellul, *Betrayal of the West*, 166.
71. Ellul, *Technological Society*, 387–427.
72. Ellul, *New Demons*.

The motto is: the more we consume (if only vicariously) sex and violence, the more technological objects we will consume; the more objects we consume, the more instinctual power we possess.

Caillois observes that excess is at the "heart of the festival." The more one pushes the negative pole, the greater the abundance at the positive pole. In the milieu of technology the festival is primarily centered in the excess of sex and violence in the mass media. As the ritualization of the principle of re-generation, the festival always begins with the negative pole—death, evil, the instinctual (sex and violence)—in the movement to the positive pole—life, good, efficiency (abundance of technological objects).[73]

In the milieu of technology, society is still sacred but in a second-ary way. Societal power remains formidable and is organized around the poles of the nation-state and revolution.[74] The technologization of society aggravates political problems by reducing politics purely to a matter of power. With the disappearance of hierarchy (in Dumont's sense), there is nothing left but competition for power among special interest groups only partially controlled by the political state. As de Jouvenel has clearly shown, political revolutions in the past several centuries have only led to an increase in state power.[75] Just as technology effectively operates outside the moral domain, so does politics today.

Scapegoating continues in the posthistoric period (the milieu of tech-nology), but in greatly diffused and secularized forms. Most of the conflicts over gender, ethnicity, and race, for instance, have gone beyond the point of legitimate grievances to that of scapegoating. With the serious decline of a common morality in the milieu of technology, all struggles tend to be con-flicts over power. The only serious issue today for any group that has been previously disadvantaged is the achievement of an equality of power. A *full* equality of power, of course, cannot be achieved without complete homoge-neity—the standardization Kierkegaard and Tocqueville feared.

Girard notes the decline of effective scapegoating in the modern era but attributes it largely to a gradual desacralization of the scapegoating mecha-nism.[76] This does not suggest a decline in the amount of scapegoating, which may actually increase, but rather points to its transitory character. I concur but have arrived at a different reason for this desacralization. While I agree that Scripture exposes scapegoating for what it is and that scapegoating is only effective when it remains unconscious, I think that the primary reason is to

73. Caillois, *Man and Sacred*, 98.
74. Ellul, *New Demons*, 82-87.
75. Jouvenel, *On Power*.
76. Girard, *Things Hidden*, 126.

be found in the gradual transition to the milieu of technology, beginning in the nineteenth century, which replaced scapegoating with another principle as the foundation of order. As society becomes subordinate to technology, scapegoating has less to do with order and consensus and more to do with the endless group competition for the power that actually resides in the techno-logical system and the political state (in a secondary way).

Yet politics (morality) remains on an illusory ideological level, an arena in which we attempt to resolve problems that are almost exclusively techno-logical. As we have seen, nature once provided the ideological and idealistic backdrop for society; now society does the same for technology.

At this point we must confront Girard's concept of mimetic desire. This, too, I find to be peculiar to the milieu of society. My argument is that imitation and desire, like ritual and the sacred, exist in relation to the domi-nant life-milieu. Moreover, it is only in the milieu of society that imitation and desire initially have different objects.

In the milieu of nature, humans were so immersed in and dependent upon nature that there was little if any awareness of the distinction between animal and human. As previously mentioned, the earliest cave art has animals almost exclusively as its subjects. When humans first enter the artistic scene in the Upper Paleolithic period it is in ceremonial relationship with animals, on an equal footing with them.[77] There is even some evidence that humans imitated the appearance of animals by wearing animal skins and so forth. My conjecture is that both imitation and desire were affixed to nature. Desire that has not been so decisively influenced by a social milieu that it becomes the desire to be like someone else centers on the possession and consumption of those natural objects that ensure survival and bring pleasure. Natural desire and imitation are channeled into the feast: I become what I eat.

In the milieu of society, imitation and desire are both separate and united. Girard has described the process by which the initial object of desire and the rival for this object become conflated. My desire is now the desire to imitate my rival, the obstacle to the initial object of desire. Or better yet, I desire the object of my rival's desire because I desire to be like him. The ini-tial separation of object of desire and imitation is finally overcome. Girard's great insight reinterpreted is that desire finally comes to reflect the milieu of society. Were it not for objects of desire that are distinct from the rival to begin with, there would be no imitation nor social desire. Only in the milieu of society does the individual perceive a clear distinction between nature and society, on the one hand, and society and technology, on the other hand. One can readily understand how social or mimetic desire and

77. Marshack, *Roots of Civilization*, 272.

the rivalry for status is a deadly problem. Someone must be eliminated—the scapegoat. Mimetic desire is channeled into ritual scapegoating: I become and do not become (the ambivalence toward the scapegoat who performs the "service" of restoring order) what I sacrifice.

In the milieu of technology desire and imitation are united as they were in the milieu of nature; there is no separation even initially, between the two. The technological milieu consists both of a system of rationalized methods and technological objects and the universe of visual images (especially in the media). Desire and imitation exist in relation to these, so in this milieu only technological objects and the material visual images that "represent" them are permitted.[78]

No one has understood this better than Guy Debord in *Society of the Spectacle*. The spectacle as the "image-object," the visual image of a commodity, is the main product today.[79] The spectacle is the "catalogue of commodities" intended for both real and vicarious (through visualization) consumption. The spectacle includes both commodities as visual images and material visual images of commodities. Reality is experienced in and through the spectacle. Two aspects of this require further clarification: How have humans come to regard themselves and others as commodities, consumer goods, and technological objects, and why have the material visual images of the mass media become more real than reality itself?

In *The Man Without Qualities*, Robert Musil depicts the plight of modern man as functionary. Without a moral core to the self, the human becomes a mere role-player, one who adjusts to reality on the one hand and manipulates it on the other. Technology and bureaucracy, as the major objectifying and depersonalizing forces in this milieu, reduce the human to an object of manipulation, thereby reducing individuality. But nothing really changes when one is the employer of technology, for now it denies one's subjective reason and experiences. Whether as manipulated or manipulator, the individual is emptied of subjectivity, qualities, and experiences.[80]

David Riesman once remarked that the main product in demand today is a personality.[81] We are constantly told to package ourselves, to market ourselves, or to sell ourselves better. There are courses both in and out of the university to assist us in this endeavor. And then too there is the enormous proliferation of shirts advertising some product. We are one with the product we advertise. Once humans wore animal skins; now they wear shirts bearing

78. Ellul, *Technological System*, 47.

79. Debord, *Society of the Spectacle*, paragraph 15.

80. Musil, *Man Without Qualities*.

81. Riesman, *Lonely Crowd*, 46.

the picture of a product. The reified human becomes just one more product on the market. We consume one another sexually or aesthetically; we become connoisseurs of body parts and personalities.

Technology's mediation of human relationships makes life more abstract and impersonal. The visual images of the mass media, however, provide an ersatz reality often more emotionally satisfying than the reality of our own lives.[82] Reality is on television and in the movies. As I argued earlier, the visual images of the media have deconstructed symbolic language so that the images ultimately are only images of possessions (objects) or images of power, objects (human or otherwise) acting in relation to other objects—acts of possession, consumption, control, or violence. We become spectators of life, visually consuming the exciting actions of objects. These image objects that comprise the spectacle are the stuff of both desire and imitation.

Stuart Ewen has recently written about some of his college students in a course on the media who exhibited the "hunger *to be* an *image*."[83] Guy Debord argues that the individual consumer is so dispossessed of himself that he has no choice but to *imitate* what is shown him.[84] As Ellul puts it, "images are norms in a world without meaning."[85]

Not just imitation but desire as well is unleashed by technology and visual images. Marx observed that desire increases with the growth in technology. The more technological objects we see, the more our desire is titillated. Ellul observes that the enormous proliferation of visual images results in our "demand for everything immediately."[86]

Nowhere do desire and imitation come together more forcefully than in the celebrity. The spectacle of the human is a definition of the celebrity. Debord dissects the celebrity:

> The agent of the spectacle placed on stage as a star is the opposite of the individual, the enemy of the individual in himself as well as in others. Passing into the spectacle as a model for identification, the agent renounces all autonomous qualities in order to identify himself with the general law of obedience to the course of things. The consumption celebrity superficially represents different types of personality and shows each of these

82. Ellul, *Humiliation of the Word*.
83. Ewen, *All Consuming Images*, 6.
84. Debord, *Society of the Spectacle*, paragraph 219.
85. Ellul, *Humiliation of the Word*, 147.
86. Ellul, *Humiliation of the Word*, 209.

types having equal access to the totality of consumption and finding similar happiness there.[87]

The celebrity is a caricature of the human, a visual stereotype. Technical desire and imitation are channeled into experimental consumption, especially of visual images: I become what I behold and consume.

Here we see the impact of the previous milieu, since the preceding milieu not only is mediated by the subsequent milieu but also provides an ideological model for it. With the celebrity, then, we can have it both ways—in effect the celebrity is only an image-object, but one that appears to be fully human.

Mimetic desire and rivalry are an artifact of the milieu of society. Competition for survival in the milieu of nature was competition with nature itself. Because of the anticipated abundance of resources and the small size of and dispersion of human groups, conflict between humans was controlled in part by the redistribution of food. In the milieu of technology, where desire is unleashed in full force along with the imitation that accompanies it, there is savage competition for consumer goods and services, but the object of imitation is far removed in these object images of the media. The violence that ensues is not mimetic, but only a competition for scarce resources. Technology and the political state have only been partially successful in controlling this competition.

The festival is the occasion for fulfillment of desire and the regeneration of the milieu. It is sacred time, and as such it produces what Victor Turner calls *communitas*, the sense of being all in one and one in all that comes from excess and renewal, that which is out of the ordinary.[88] There is no single act or principle of origin that can explain the festival, for it is relative to the human life-milieu.

87. Debord, *Society of the Spectacle*, paragraph 61.
88. Turner, *Ritual Process*.

Chapter 11 _____

The Legend of "Nigger" Lake:
Place as Scapegoat

Originally published in the *Journal of Black Studies* 28, no. 6 (1998) 704–23.

Lauren E. Brown, Emeritus Professor of Vertebrate Zoology,
Illinois State University, was the coauthor.

The term *scapegoating* has had three principal meanings that have not always been clearly differentiated: biblical, anthropological, and psychosocial. Originally, the scapegoat was a term used to refer to the first of two goats in the Leviticus ritual, the one chosen to carry (symbolically) the sins of the Hebrews into exile.

In the nineteenth century, anthropologists generalized from the biblical ritual and applied the term to a wide range of rituals, which later became a subcategory called scapegoat rituals or rites for the expulsion of evil. For at least twenty years, however, the anthropological use of this category has been moribund.

Finally, in everyday discourse, in journalism, and in popular literature, *scapegoat* is a term used to refer to those unfairly blamed for problems of any kind. This general psychosocial meaning has become predominate.[1]

Despite the fact that the psychosocial meaning of scapegoat is widespread, the term has been given short shrift in the social sciences. Two approaches characterize those who do not use scapegoat in an exclusively descriptive way. Some social scientists, often psychologists, have attempted to account for the motivation to scapegoat; others, usually sociologists, have examined the consequences or functions of scapegoating.

The typical theory of the motivation for scapegoating is the frustration/aggression theory. Here, frustration within the in-group, whether due

---

1. Girard, "Generative Scapegoating," 73–74.

to internal conflicts or stress, creates a need for aggressive behavior, which in turn is displaced onto a member of or the entire out-group. Gordon Allport observes that frustration does not always lead to aggression, that aggression does not invariably result in displacement, and that scapegoating cannot account for all forms of prejudice.[2] Still, he devoted considerable attention to it. The famous study *The Authoritarian Personality* used the frustration/aggression theory to explain scapegoating.[3] Allport and Adorno, et al., were instrumental in briefly bringing scapegoating before professional audiences and the public in the mid-twentieth century.

The consequences of scapegoating, however, have rarely been studied in part because they appeared obvious—an increase in hatred of and prejudice against those scapegoated. Or when the consequences have been studied, they are sometimes set within the general process of labeling someone deviant.

In a number of books on mental illness, for example, but especially in *The Manufacture of Madness*, Thomas Szasz analyzed the mental patient as scapegoat.[4] He accepted frustration as the motivation for scapegoating, concentrating instead on its functions, which included group solidarity, a sense of superiority, and avoidance of one's own involvement in evil.[5] Szasz explicitly made scapegoating a category of labeling within the deviance perspective in sociology.[6]

But what if ritual scapegoating is the origin of culture and religion itself? And what if societal consensus rests on the choice of a common victim, the scapegoat? French literary critic turned social scientist, Rene Girard, has made these arguments and offered them as scientific hypotheses.[7] If he is on the right track, the assumptions of the social sciences about social order will need to be reconstituted. His work, then, has enormous implications for sociology as a whole. Girard's position is the opposite of the functionalist argument: first social order based on consensus, then scapegoating to reinforce that order. He argues that scapegoating brings about social order and its consensus. His theory of mimetic desire and mimetic rivalry on the psychological level is an alternative to the frustration/aggression argument and is quite specific about the interpersonal conflict that results in collective scapegoating. Finally, he demonstrates how ritualized scapegoating

2. Allport, *Nature of Prejudice*, 325–33.

3. Adorno et al., *Authoritarian Personality*.

4. Szasz, *Manufacture of Madness*.

5. Szasz, *Manufacture of Madness*, 260–75.

6. Szasz, *Manufacture of Madness*, 279–80.

7. Girard, *Scapegoat*; Girard, "Generative Scapegoating"; Girard, *Things Hidden*.

reinforces the extant order and helps to militate against the interpersonal conflict that gave rise to the scapegoating in the first place.

It is not our aim to test his larger theory or even to argue its greater merits as an explanation of scapegoating.[8] Our main interest is in his description of the three-step process by which a scapegoat is selected. By contrast, Allport had earlier argued that the choice of scapegoat simply varied with historical circumstances. One can accept Girard's theory of the scapegoat selection process, however, and still leave open the question of his larger theory.

Racism and scapegoating seem to go together: the scapegoat is sometimes a feared and detested race, just as racism involves blaming the other and denying one's own responsibility. Scapegoats are almost invariably imagined to be people or, on occasion, animals, but rarely as geographical places. We argue that a place can act as a scapegoat when it becomes a symbol of the outsider, who is human. In this article, we wish to consider how a lake, symbolic of the African American as outsider, became a scapegoat for a community's flooding problems.

The African American Grocer

About 135 years ago, an African American male operated a small grocery store just south of an ephemeral lake (see Figure 1) in the sand prairie southeast of Havana in Mason County, Illinois.[9] Why he selected this locality is unclear, but legend suggests that he was a driver of a circus wagon that became mired down in the road crossing the wetland. We suspect he avoided living in Havana (a river town) to escape racial prejudice and harassment. There is historical evidence that he both served whiskey and permitted card playing in the back of the grocery (this was not atypical at the time). Eventually, the wetland became known as "Nigger" Lake by local people, and it is still commonly referred to by that name. In 1993, local officials, who were applying for grants to alleviate the flooding problem, gave the wetland the name Sand Lake to avoid public embarrassment.[10]

8. Stivers, "Festival in Light," ch. 10 in this book.
9. "Old Stories about Negro Lake"; Speckman, "Facts about 'Nigger' Lake."
10. W. Ingram, personal communication (1996).

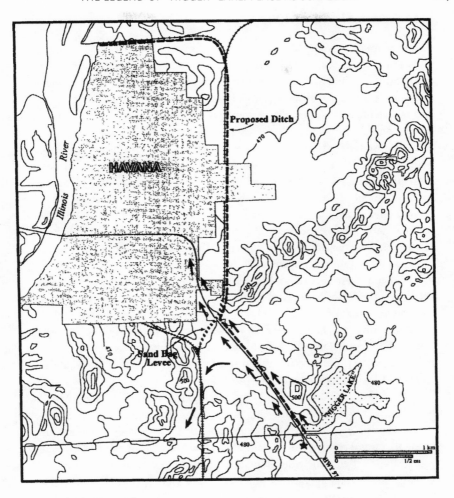

Figure 1: Map of "Nigger" Lake and Havana, Mason County, Illinois

SOURCE: Adapted from maps prepared by Environmental
Science and Engineering and the U.S. Geological Services (1948).
Cartographer: Jill Freund Thomas.

NOTE: Darkened star indicates approximate location of grocery
store operated by African American male; arrows indicate direction
of water flow from "Nigger" Lake into Havana; heavy dashed line
indicates proposed location of ditch to drain the wetland.

The Great Flood of 1993

At approximately twenty to forty year intervals (1888 [1898?], 1926 to 1927, early 1940s to 1946, 1972 to 1974, 1993 to 1994) "Nigger" Lake experienced substantial flooding and expansion much beyond its usual size, with over-flow into Havana. A wet spring and summer in 1993 caused wet basements and failure of septic systems in certain areas of east Havana. This foreshad-owed far greater problems when, in September 1993, two particularly heavy rainfalls resulted in the area's worst flooding in recorded history. The conse-quence was more than 171 groundwater lakes in Mason County, including a greatly expanded "Nigger" Lake.[11] The latter overflowed northward, via roadside ditches (see Figure 1), inundating east Havana. Residential base-ments suffered significant structural damage, and many homes were nearly uninhabitable. Two schools as well as a number of commercial and industri-al businesses were also adversely affected. Therefore, the entire community was directly or indirectly harmed by the flooding.

Citizens were frustrated, frightened, and sometimes angry; moreover, the large number of haphazard private meetings to discuss the flood sug-gests a community in a state of chaos. Eventually, a series of well-attended public meetings was held, and many individuals demanded with no opposi-tion that "Nigger" Lake be drained. On at least three occasions at public meetings, several individuals argued without explanation that the water in "Nigger" Lake had come from Ohio and New York.[12]

Environmental Science and Engineering, Inc. was contracted to ana-lyze the flooding problem and suggest solutions. The cause of the flooding in the Havana area was a high water table that had risen above ground level. This was a result of heavy rainfall, supersaturated sandy soils, limited topographic relief, poorly developed natural drainage systems, subsurface geology, and probably various other factors.[13] Hence, the "Nigger" Lake wetland was a symptom and not a cause of the flooding. Consequently, W. Ingram, Water Resources Engineer of Environmental Science and Engineering, pointed out that the drainage ditch would not alleviate, to a substantial degree, the flooded basements or other flooding problems unrelated to "Nigger" Lake.[14] (Recall that east Havana was already having water problems in the summer of 1993 before the overflow of "Nigger" Lake in September 1993.) It was consequently questionable if the ditch

11. Clark, *Mouth of the Mahomet*.

12. W. Ingram, personal communication (1997).

13. See Brown and Cima, "Illinois Chorus Frog"; Clark, *Mouth of the Mahomet*.

14. Williams, "Ditch Proposed."

would provide adequate relief to Havana. Furthermore, preliminary cost analysis by Environmental Science and Engineering indicated that the benefits of the ditch are only about equal to the construction costs. Many local residents were unconvinced that "Nigger" Lake was not the cause of all of their water problems, in spite of repeated attempts of Environmental Science and Engineering to enlighten the community.[15] The public was obviously preoccupied with draining (and hence destroying) "Nigger" Lake, even though there was no rational reason to do so.

Environmental Science and Engineering suggested three short-term options that were implemented: sand bagging, pumping, and ditching. Although relief was not immediate, these measures (plus evaporation) eventually eliminated the flooding problem. Environmental Science and Engineering also identified several long-term options that might help relieve future flooding. A drainage ditch extending from "Nigger" Lake along the east side of Havana and then into the Illinois River (see Figure 1) was most popular among local citizens. This plan was favored because it was perceived as a solution for draining "Nigger" Lake, and because it might also provide a means for draining east Havana.[16] The ditch was estimated to cost $1.3 million. Efforts to obtain funding have so far been unsuccessful, but citizen concerns remain high, and planning continues.[17]

For us, there is no solution to potential flooding given the extremely high water table of the area. The causes of the flooding were the large amount of rainfall and high water table combined with inadequate natural drainage. When there is no true solution to a problem, a group sometimes looks to a false solution as a way of reuniting the community.

Racism In Havana

Havana is a large small town with a population (Bureau of the Census, 1990) of slightly more than 3,600. It is primarily a working-class community with four African American residents (one family). This is typical of towns this size in central Illinois; indeed, there are only eight African Americans in Mason County, which has a population of about 16,000. Havana is a former port town and, as such, may have witnessed a number of African Americans pass through over the years.

There are two kinds of racism: one that is direct and material, a result of a fear of and a desire to subjugate the other race's potential physical,

15. W. Ingram, personal communication (1997).
16. W. Ingram, personal communication (1997).
17. Brown and Cima, "Illinois Chorus Frog."

political, or economic power; the other, indirect and symbolic, a consequence of a fear of the other race's standing as a degraded, inferior group. The two racisms are intimately connected.

It is indisputable, we think, that racism against African Americans has been virtually universal in American society. Alexis de Tocqueville observed in the 1830s that the racism in the South was more direct—institutionalized in slavery and later in segregation—than that in the North.[18] Nevertheless, Northern racism was not weaker; it was merely different. His argument was that as long as there are legal and other institutional barriers separating the races, the dominant race feels secure to establish informal relationships with the exploited race. The dominant race has nothing to fear, either materially or symbolically. When there are fewer institutional barriers between the races, as in the North, the dominant race is more fearful—less fearful of the economic and physical power of the exploited race and more so of being contaminated by a socially degraded, disreputable race. In the South, the relations between the races worked as long as African Americans knew their place and stayed within its boundaries. When African Americans deviated from Southern norms, the dominant race often expressed violent racist attitudes. Northern racism, although it was always capable of surfacing explosively, was often expressed in more subtle ways, such as covert discrimination, avoidance, and denial.

The African American was a symbol of degradation intensified by the relative lack of structural barriers in the North. The fact that some Northern residents (such as in Havana) rarely came in contact with African Americans does not diminish the fear and repugnance, because the latter are magnified by the lack of barriers between the races, not by the frequency of interaction. There is an inverse relationship, it appears, between structural barriers of discrimination and the intensity of fear of the dominated race as a symbol of degradation. The racist logic is that social contact with a so-called inferior race without strict regulations can lead one to catch degradation as a kind of contagious disease.

Havana is by no means exceptional. We are not suggesting that it is more racist than any of its neighboring towns. But the facts indicate that there is racism here. One native informant, a white female in her twenties, remembers that as a child she played a game called "Nigger Knocking," which involved several children knocking on someone's front door and running away to hide before the occupant could answer the door. The informant maintains that the game was widespread in Havana but does not know the origin of the game's name. She also claims that the road running near

18. Tocqueville, *Democracy in America*, 316–407.

"Nigger" Lake was commonly referred to as "Nigger" Lake Road. She insists that Havana is a racist community.

A second native informant (the young woman's aunt), a white female in her forties, disagrees about racism in Havana. She maintains that Havana cannot be racist because it has an ordinance prohibiting African Americans from living there. (Whether Havana ever had such an ordinance, it likely had an informal norm to the same effect.) The racist logic is that one cannot be a racist if there is no one nearby to discriminate against.

Even without the observations of the informants, the name "Nigger" Lake and the legend about the "nigger" who operated a grocery near it are evidence that African Americans are still symbolic of contamination and degradation. There need not have been lynchings, beatings, or harassment for there still to have been racism in the town. Havana provides evidence of an indirect or symbolic racism widespread in American society.

Symbolism of Place: The Outsider

Keep in mind that the wetland ("Nigger" Lake) is geographically outside Havana (approximately one mile to the southeast of the city); therefore, the citizens do not regularly encounter the lake in their daily activities. The concept of outside, of course, can refer to more than physical space; it often is applied to social space as well. What is outside of our social space is different, foreign, and potentially dangerous to our way of life.

The history of comparative religion has long recognized that place has a religious meaning in traditional societies, which in part derives from an opposition between "our world," the place we inhabit, and a "foreign, chaotic space," peopled by outsiders, foreign spirits, and even demons.[19] Our world or place, whether a nomadic territory or a village, is sacred; by contrast, foreign space is profane, that is, without ultimate meaning. But it is not just a matter of intellectual contrast, that is, profane space versus sacred place. Profane place, the people who inhabit it, and its values threaten to desecrate the sacred place and its inhabitants. The main reason is that sacred space is the place where (our) nature or society was created. And creation is equivalent to good, just as the threat to creation is evil.

A symbol, whether a word, phrase, object, or event, possesses indirect meaning: It stands for something else. All symbols beyond their literal meaning have a figurative meaning. Water, for example, appears to be an almost universal symbol of fertility, birth, and rebirth. As evidenced by our example, symbols have multiple meanings that are nonetheless

19. Eliade, *Sacred and the Profane*, 29–32.

interconnected.[20] The indirect or figurative meanings of a symbol are formed by the comparisons and associations we make. "Nigger" Lake, for instance, is a lake that symbolizes the outsider, specifically, the African American, perhaps the most feared and detested outsider in American history. The lake or wetland has periodically invaded and polluted the physical structure of community, just as African Americans, it is imagined, can desecrate the social structure of the community.

Nigger is the master symbol of outsider to the community, we surmise, but there are additional secondary symbols. The master symbol organizes and gives further meaning to the secondary symbols, which in turn imply the former; the master and secondary symbols thus form a complex.[21]

Birders (amateur ornithologists) from other areas of the state frequently visit the wetland to observe the abundant avifauna. These individuals exhibit unusual behavior (they stand in a stationary position for long periods of time while looking through binoculars) and often wear unusual attire (strange hats and floppy jackets to hold field guides, snacks, etc.). Waterfowl are exceedingly abundant at the wetland, and the area has been attractive to hunters, as evidenced by the presence of duck blinds in the wetland. The area along the Illinois River has been of major importance to duck hunters for more than a century.[22] Many duck clubs have been established, and numerous hunters from Chicago are attracted to the area.[23] These sportsmen carry guns (a potential threat), talk faster, and have a different (foreign) accent. Speckman pointed out that whiskey drinking and card playing, which was probably accompanied by gambling, occurred at the original grocery store owned by the African American some 135 years ago.[24] (On two separate occasions, local residents have communicated the same information to L. E. Brown [one referred to the store as a *tavern*].) Thus, the grocery store could probably be viewed in the same light as a roadhouse. Such establishments often had extremely bad reputations because they offered bootleg liquor, gambling, cock fighting, dog fighting, and so forth.[25] Because roadhouses were located in rural areas, they were unregulated by village ordinances. Finally, some Havana residents voiced the preposterous belief that the local flood waters originally came from Ohio and New York. Supposedly, floods that occur in the eastern United States result in flooding at "Nigger" Lake and Havana at a

20. Eliade, *Patterns in Comparative Religion*.
21. Ricoeur, *Interpretation Theory*.
22. Parmalee and Loomis, *Decoy Carvers of Illinois*.
23. Parmalee and Loomis, *Decoy Carvers of Illinois*.
24. Speckman, "Facts about 'Nigger' Lake."
25. Angle, *Bloody Williamson*.

later date.[26] Thus, there is substantial evidence that "Nigger" Lake has outside (foreign) characteristics that are disapproved of or feared by local residents of Havana. In summary, the lake symbolizes *Nigger*, master symbol of the outsider. In addition, the lake symbolizes bird watcher, wealthy duck hunter, roadhouse, and eastern United States (people in the Midwest often look on the East Coast as a problem-laden and unfriendly area).

We are not suggesting that simply calling the wetlands "Nigger" Lake would have been sufficient to bring about its status as scapegoat. Rather, it was all the ways the lake was used by outsiders in the present that allowed the symbol of nigger from the past to be inexorably connected to them. Symbolic connections are not logical but associational: Nigger equals wealthy duck hunter, eccentric birdwatcher, and so forth. At the same time, however, the hostility toward outsiders confirmed the latent racism in the community. The various associations among the meanings of a symbol are mutually reinforcing.[27]

Ritual Scapegoating

"Nigger" Lake is more than a symbol of the outsider, however; it is a ritual scapegoat. Although the concept of scapegoat was used in academic life as early as the nineteenth century, no one had satisfactorily worked out its larger theory until Girard discovered a connection between mimetic desire and ritual scapegoating.

Mimetic desire is desire complicated by competition for the object of desire. Desire becomes mimetic when one desires what someone else desires; the other becomes a model for me: I desire to be like the person who desires what I now desire. I am thus attracted to this other person. Concurrently, however, this other is a rival: we both desire the same object. The object of desire grows in importance as the competition for it intensifies.

My rival gives two messages: Be like me and do not be like me. The rival is flattered at first by imitation because everyone wants to be admired; the initial feeling of pride is quickly countered by the fear of losing the object of common desire. As a rival, the other is an obstacle to the realization of desire. Hence, I am repelled by my rival. Attraction and repulsion.

Mimetic rivalry is contagious, as is the violence it engenders. Violence is itself mimetic. I respond to your violent act with a similar act of my own. We become like each other in our violence. Still others enter the fray, because

26. W. Ingram, personal communication (1997).
27. Ricoeur, *Interpretation Theory.*

each momentary victory attracts new opponents. This is the mimetic crisis: the certain prospect of widespread violence with no apparent solution.

A solution emerges spontaneously, one that is the originary act from which ritual, religion, and society itself spring forth. The solution is a collective murder that both ends the violence and symbolically unties the remaining competitors—the killing of a common enemy, a scapegoat.[28] What may appear as an arbitrary act is not recognized as such by the collaborators; the scapegoat is not recognized as scapegoat. For the act of ritual killing to be an effective agent of unification, it must not be recognized as an act of scapegoating. The scapegoat will be viewed as the source of all evil and, hence, as one who deserves to be killed. The murder of the scapegoat unites the members of the society by permitting evil to be expelled. Society cannot confront the arbitrary nature of its violence and of its own complicity in the contagious violence preceding the original act of scapegoating. Therefore, religious myths conceal the act of murder as scapegoating; instead, they represent the murdered as fully deserving to be killed. Simultaneously, however, the ritual victim is sacralized, transformed into a deity who now can be made to work for the benefit of society. For instance, in the Seneca myth of creation, after the good twin (the Creator) defeats his brother the evil twin (the Great World Rim Dweller), a pact is made between the evil twin and the Seneca.[29] In return for food and supplications, the evil twin would "cure incurable disease, avert deadly tornadoes, cast out malevolent witches, and bring order to a whole community."[30]

The originary act of murder is not just kept alive in myth; it is systematically repeated in ritual. The reenactment is often a dramatic enactment rather than an actual repetition of the originary act of killing. There may likewise be a substitution for the ritual victim—animals, even crops, may become symbolic surrogates. In Girard's view then, the ambiguity of sacred value resides in this fundamental contradiction: violence (evil) is the origin of order (good).[31] If violence is sacred, it is so only because it is the source of unity, the order of all against one.

Ritual scapegoating in the modern era, however, does not deify the victim. Girard comments on this in the following passage.

> Religious phenomena are essentially characterized by the double transference, the aggressive transference followed by the reconciliatory transference. The reconciliatory transference

28. Girard, *Scapegoat*.
29. Wallace, *Death and Rebirth*.
30. Wallace, *Death and Rebirth*, 92.
31. Girard, *Things Hidden*, 32.

sacralized the victim and is the one most fragile, most easily lost, since to all evidence it does not occur until the mechanism has completely "played itself out." We remain capable, in other words, of hating our victims; we are no longer capable of worshiping them.[32]

The farther away from its violent origins a community and its religion get, the more ritual scapegoating loses its meaning and purpose. The time comes when scapegoating will be justified for political reasons.[33] As scapegoating becomes political (secular), it becomes more widespread but simultaneously shallower. That is, scapegoating today can still generate hatred, but it is now directed toward a plethora of transitory victims.

What is most important for our purposes is Girard's account of the dynamics of selecting a scapegoat: "stereotypes of persecution."[34] Scapegoating is actually precipitated by two crises; the second crisis, the more serious one, is a comment on the first. The initial crisis can be almost anything: famine, military defeat or victory, political conflict, hurricane, flood. But the real crisis, in the social sense, is the reaction to the first one. This reaction, whether violence, frustration, anger, or even ecstasy, tends to be uniform, which means that social differences have been momentarily eliminated. As Girard observes, every society is organized so that there is both unity and diversity; the latter is in fact a necessary component of the former. That is, customs, mores, manners, and behavior differ by class, sex, age, and so forth but exist within a larger culture that permits such diversity. The real social crisis is that virtually everyone acts the same in an emergency—hoarding food, looting, becoming violent or angry—and, in so doing, denies the extant social structure.[35]

The reaction to the social crisis of homogeneity involves making an accusation and choosing a victim. The scapegoaters accuse an individual or group of some moral failure that caused the initial crisis. For example, during the fourteenth century, the Jews were accused of poisoning the rivers in northern France, which in turn resulted in the famous Black Death or Plague (the first crisis). The second crisis was the homogeneous way people reacted to the Plague—with suspicion, isolation, distrust, generalized anger, and violence. Often, the accusation is about an immoral act, in this instance poisoning, one of the most heinous crimes in traditional society.

32. Girard, *Things Hidden*, 37.
33. Girard, *Scapegoat*, 113.
34. Girard, *Scapegoat*, 12–23.
35. Girard, *Scapegoat*.

Finally, Girard notes that there are universal signs for the choice of scapegoat. The victims are always those who are different, whether ethnically, culturally, or physically; moreover, these differences must be observable. The victims are chosen from among those who are marginal to society. In the above example, the Jews, who were marginal to Christian society, were the scapegoats. The victim or scapegoat is believed to be responsible for the first crisis, the Plague, but secretly the scapegoat is thought to be responsible for the second (social) crisis—the lack of differences in attitude and behavior.[36]

Here is the crux of the scapegoating mechanism. By identifying a victim as the cause of the cause of disorder in society, as responsible for the original crisis (famine, plague, flood) that, in turn, leads to the crisis of contagious violent behavior, the scapegoaters are simultaneously establishing boundaries between the system and what lies outside and reestablishing diversity of action within the system. The scapegoat belongs outside the system because he or she caused the individuals within the system to shed their diversity and become dangerously alike.

"Nigger" Lake As Scapegoat

The narrative of how local residents reacted to the Great Flood of 1993 perfectly fits Girard's theoretical description of the scapegoating mechanism: (a) social crisis, (b) accusation, and (c) choice of victim.

The residents of Havana experienced residential basement flooding, and several schools and businesses suffered flood-related problems; this is the first crisis. Subsequently, they began to act similarly—wildly uncertain with fear, frustration, and anger, they expressed hostile attitudes toward "Nigger" Lake; this is the second crisis.

Accusation and choice of victim are intimately related. One might argue that the accusation is framed in terms of what is commonly believed the accused (scapegoat) is capable of doing. The moral accusation against the wetland is that it deliberately polluted (flooded) parts of the community. We say deliberately because in refusing to listen to the scientific evidence that "Nigger" Lake was not responsible and in wanting it destroyed, the citizens acted as though they believed the lake was personally out to get them. And is this not because the lake symbolizes the outsider, principally the African American, the quintessential marginal group in American life? The lake is a surrogate scapegoat, the outsider poised to destroy the physical community only because, it is believed, African Americans have always been ready to

36. Girard, *Scapegoat*.

pollute the social community. The citizens attempted to restore unity to their community by finding a common scapegoat. Although there was no readily believable human victim, a place with a history of marginality, "Nigger" Lake, was a convenient scapegoat. But it was a convenient scapegoat in large part due to its symbolic meaning. Scapegoating appears, unfortunately, to be an integral part of the social order, and it extends even to place.

The Basic Data And Their Fit With Scapegoating

We have relied on the following sources of data: the corroborated memories of an environmental engineer who attended the public meetings about the flood as a professional consultant; the historical record of newspaper stories about "Nigger" Lake; interviews with two native informants; and the observations that one of the authors, a biologist studying the wetlands, made about the usage of "Nigger" Lake. We examined twenty-five stories about the flood of 1993 in three different newspapers, but none of them revealed details about the content of the public meetings. Because we did not begin this study until 1995, we have neither participant observation data from the various meetings nor interviews with residents.

Yet, obviously it is not the quantity of data that matters most but the relevance of the data. The views that were spontaneously expressed at public meetings on the flood and in the aftermath are much better indicators of real sentiment than those that would have been expressed in an interview or questionnaire at a much later date. Moreover, we have the basic data necessary for an interpretation:

1. the view at the public meetings, uncontested and seemingly unanimous, that the cause of the flooding problems was "Nigger" Lake, despite expert opinion to the contrary;

2. the view, at the public meetings less widespread but still uncontested, that much of the water in "Nigger" Lake came from Ohio and New York; and

3. the various kinds of outsiders (as observed by one of the authors) who dominate the use of "Nigger" Lake.

Because the wetlands popularly called "Nigger" Lake did contribute water to the flood of 1993, the townspeople's attitudes about the wetlands causing the flood are seemingly rational, but the first two data taken together suggest that the dominant attitudes toward "Nigger" Lake are irrational. The last two data suggest that "Nigger" Lake is associated with outsiders (the water and its use). The second datum combines the irrational with xenophobia (fear of outsiders). The fear and hatred of water from the eastern United States is the key to understanding the other two data and the larger mystery of "Nigger" Lake. It demonstrates that the irrational attitudes are directed toward outsiders and that, because the attitudes toward outsiders are irrational, the outsiders symbolize evil. Scapegoating possesses all these characteristics.

Everyone who has studied scapegoating agrees that it is an unconscious, symbolic process. No one who scapegoats believes anything but that the scapegoat is a deserving object of rejection. We contend that indirect data, such as what we possess, are for certain issues, especially symbolic ones, much more important data than the direct data of interview and questionnaire. Keep in mind that shared symbolic meanings are indirect meanings that often arise spontaneously and unconsciously and that their partly unconscious character gives them their efficacy.[37] Our interpretation, which relies on the unconscious process of scapegoating, fits the data, but are there superior interpretations?

Alternative Interpretations

We have offered a plausible interpretation of the attitudes of some of the residents of Havana toward "Nigger" Lake. But there are more obvious interpretations. Could it be that the residents of Havana who attended the town meetings at which the causes of the flooding were discussed did not understand the engineers' arguments, perhaps were even hostile to science, or simply preferred the evidence provided by common sense? Given the fact that the scientific explanations were couched in commonsense terms and were straightforward—high water table, poor drainage, and heavy rainfall—it is unlikely that they did not understand the engineers. It is improbable that they were antiscientific when they invited the engineers there in the first place. Attribution theory in psychology might suggest that there are discernible reasons other than scapegoating for why individuals make mistakes in perception about their environment. Attribution theory works on the psychological level, scapegoating theory on the sociological level. Fritz Heider, one of the leading attribution theorists, states that sometimes the "affective significance"[38] of an object or event can determine the attribution of causes, in this instance, the cause of the flood. When affect overwhelms intellect, attribution is a variant of rationalization. Attribution theory here

---

37. Ricoeur, *Interpretation Theory.*
38. Heider, *Psychology of Interpersonal Relations*, 171.

actually supports our interpretation because townspeople can always rationalize their unconscious rejection of "Nigger" Lake when confronted by contrary evidence. Their rationalization is that the water that overflowed "Nigger" Lake caused the flooding problem.

Racism and scapegoating reflect, in large part, collective, irrational attitudes. No one admits to being a racist or a scapegoater. The victim always deserves what is done to him or her. Irrational beliefs are not susceptible to persuasion, at least by itself. A serious change in existential attitude demands a radical experience that runs contrary to the extant attitude.

To our mind, the most telling bit of evidence is the irrational attitude that the local flood waters, specifically those in "Nigger" Lake, originally came from Ohio and New York. This suggests, at a minimum, that some of the local residents felt the irrational need to blame the flood on an outsider. We have shown the multitude of ways in which "Nigger" Lake symbolizes the outsider. Therefore, the argument that "Nigger" Lake is at once a symbol of outsider and a ritual scapegoat is a more compelling interpretation than that of ignorance or antipathy to science.

Chapter 12 _____

Our Brave New World Today

Originally published in the *Bulletin of Science, Technology and Society* 30, no. 4 (2010) 247–51.

What does science fiction at its best reveal about our scientific and technological present, and moreover is it prophetic about the future? These are perennial questions. To these must be added Jacques Ellul's observation that science fiction is a compensation for a reality even bleaker than that presented in literature.[1] As such science fiction helps us adapt to a world that is better only by comparison. Let us examine Aldous Huxley's fictional *Brave New World* (written in 1931, published in 1932) and his non-fictional *Brave New World Revisited* (1958) as a way of providing a partial answer to questions about the significance of science fiction.

Brave New World was published in 1932; *Brave New World Revisited* in 1958. This is the only example that I am aware of in which an author later evaluates in a nonfiction work what he had earlier presented in fictional terms. In other words, he asks how *Brave New World* stands up twenty-six years later. We will ask how both his fictional and nonfictional accounts fit our world today. Perhaps in answering this question, we will better understand the significance of science fiction.

Aldous Huxley's *Brave New World* (1946) and George Orwell's *1984* (1949) are arguably the two most important works of science fiction in the twentieth century. And yet they couldn't be more different. Orwell's world is that of political totalitarianism—Nazism, Fascism, Stalinism—which inspired his narrative. Orwell's brilliance is most obvious in his depiction of the totalitarian attack on books and language.

Huxley's world is not that of a police state able to punish almost any discretion and extinguish any semblance of privacy. His is a world of pleasurable reward for conformity—a world of therapy and manipulation

---

1. Ellul, *Technological System*, 112.

instead of punishment and threat. Huxley portrays a totalitarian society that is based on scientific technique rather than politics, or more precisely, a politics that is determined by technique. Technological totalitarianism has superseded political totalitarianism.

Huxley's concern was freedom in the face of universal applied science (technique) used on humans. Orwell's paramount concern as well was freedom, which could be restored by political radicalism. Huxley's narrative is less hopeful because science and technology appear to be neutral and thus more difficult to mount an attack against.

In his 1958 *Brave New World Revisited*, Huxley maintained that his forecasts in *Brave New World* had come to fruition much sooner than he had anticipated. Huxley, who read French, was instrumental in getting Jacques Ellul's 1954 book on technology translated into English. More than likely Ellul's *The Technological Society* (1964) influenced Huxley's *Brave New World Revisited* (1958). I will employ Ellul's work on technology as the standard for understanding technique's impact on society. But first let us examine *Brave New World* (1946) and *Brave New World Revisited* (1958).

Brave New World (1931)

In the foreword to the 1946 edition of *Brave New World*, Huxley stated that "inefficiency is the sin against the Holy Spirit" in a technological society, in other words, the unpardonable offense.[2] Efficiency is required in all areas in life—both in regard to nature and to humans. Huxley understood what relatively few observers have, that technology is both material and nonmaterial. Indeed it is necessary that humans be brought into conformity with the movement of a technological civilization through the application of psychological and organizational techniques. The worldwide utopia of *Brave New World* (1946) is governed by human techniques (both material and nonmaterial).

Genetic engineering created a human caste system: Alphas, Betas, Gammas, Deltas, and Epsilons. The most intelligent type, Alphas, were engineered for the top leadership positions, whereas the Gammas for the most pedestrian jobs. The remaining types were programmed for increasingly less intellectual tasks as one moves from Betas to Gammas to Deltas. In *Brave New World* (1946) pregnancy was mechanized, thereby eliminating motherhood. The government controlled population at two billion worldwide. The ultimate material technique of human control was soma, a drug that produced a mystical euphoria and temporarily eliminated ambition and pride. A smaller dose

2. Huxley, *Brave New World*, rev. ed., xii.

of soma made one blissful, a larger does led to pleasurable hallucinations, and a maximum dose resulted in a deep and refreshing sleep. The drug had no negative side effects. It was the perfect technique—a political form of control that was simultaneously desired by the citizens.

Forced consumption was both a material and nonmaterial technique, for it depended on both technological objects and services and on the psychological techniques of hypnopaedia and propaganda. The former includes both sleep conditioning and suggestion during hypnosis. Given that children were socialized in government-run camps, hypnopaedia was readily practiced. But even more important was propaganda, which included both political propaganda and advertising. In a totalitarian system the two forms of propaganda merge. The inhabitants of Huxley's world were continuously bombarded with propaganda; there was no reprieve.

Propaganda stifles critical reflection as does its complement—the peer group. No one was permitted to be an individual—it was a taboo. Everyone was made part of a therapeutic group, which included chanting and the worship of Ford, the founder of the world government. Just as there was forced consumerism there was forced sociability. Everyone had been conditioned to look for signs of nonconformity. Privacy was an anachronism. The peer group was supreme in the absence of marriage and the family. But romantic relationships were censored: you were denied an exclusive sexual relationship. And friendships could not be close.

In the absence of intense ethical relationships, passion lay dormant. Søren Kierkegaard argued that passion follows from the exercise of moral authority.[3] For example, the reciprocal demands of father and son create strong emotions—admiration, love, hatred, fear. In Huxley's world all relationships were characterized by shallow and transitory feelings. This of course is why the government banned marriage, family, and friendship. Moreover, morality and religion (except for the worship of Ford) had been eliminated. Government destroyed any institution that might compete for the citizen's allegiance.

Even culture could not be permitted: no traditions or old books were allowed; only technical knowledge and factual information were in print. Replacing a complex culture that assisted people in confronting suffering and death and exploring the meaning of existence was a culture of happiness. But happiness was a jealous value; it permitted no others. It did so by redefining other values as simply expressions of itself. For instance, freedom was only another name for happiness.

The malcontent Bernard says, "Don't you wish you were free, Lenina?"

---

3. Kierkegaard, *Present Age*.

I don't know what you mean. I am free. Free to have the most wonderful time. Everybody's happy nowadays.[4]

In *Brave New World* (1946) Huxley gives us a story of a world utopia that forces people to be happy through various scientific techniques of manipulation and control. It is a world of control in and through pleasure. Real freedom is abandoned for it invariably brings conflict, anxiety, and suffering. This technological totalitarianism has a happy face.

Brave New World Revisited (1958)

Twenty-seven years later, Huxley (1958) examined the modern world in terms of his earlier novel, and was shocked to discover that fiction was becoming reality much sooner than he had originally anticipated. *Brave New World Revisited* (1958) is a sociological portrait of modern societies. Chapters include a discussion of overpopulation, overorganization, chemical manipulation (pharmaceuticals), propaganda, advertising, hypnopaedia, subliminal perception, and brainwashing. Most of these topics were dramatically portrayed in the novel.

Of special note are his chapters on propaganda in a democracy and overorganization. Huxley contrasts the irrational propaganda of a dictatorship with the rational propaganda of a democracy based on the enlightened self-interest of both leader and citizen. The mass communications industry threatens democracy by providing a plethora of distractions that are symbolically related to desires and fears. Therefore, the mass communications industry is not concerned with truth and falsehood but with possible imaginary realities. Democratic citizens thus have increasingly unrealistic expectations and embrace illusion. Huxley is aware that advertising and the mass media threaten a democracy as much as authoritarian control.

By overorganization he means the quest for absolute efficiency and perfect order in big government and big business and the resultant emphasis on conformity. Technology is first and foremost about the rational organization of things and people—machines, bureaucracy, computers. Huxley (1958) puts it this way:

> During the past century the successive advances in technology have been accompanied by corresponding advances in organization. Complicated machinery has had to be matched by complicated social arrangements, designed to work as smoothly and efficiently as the new instruments of production. In order to fit

4. Huxley, *Brave New World*, rev. ed., 61.

into these organizations, individuals have had to deny their native diversity and conform to a standard pattern, have had to do their best to become automata.[5]

Modern societies of 1958 therefore closely resemble the social world of *Brave New World* (1946) with several notable exceptions. First, there is no World Utopia ruled by World Controllers. Second, there is no caste system created by genetic engineering. But are these really the most important components of technological totalitarianism anyway? When psychological and managerial techniques are perfected, totalitarianism can still be achieved.

At the end of *Brave New world Revisited*, Huxley returns to freedom, his main theme:

> Meanwhile there is still some freedom left in the world. Many young people, it is true, do not seem to value freedom. But some of us still believe that, without freedom, human beings cannot become fully human and that freedom is therefore supremely valuable. Perhaps the forces that now menace freedom are too strong to be resisted for very long. It is still our duty to do whatever we can to resist them.[6]

Despite his urging, there is almost a sigh of resignation in Huxley's final paragraph.

Jacques Ellul's Sociology of Technology

We now turn to Jacques Ellul's work on technology as a way of evaluating what Huxley got right and what he omitted.

One of Ellul's critical insights is that technology involves more than material technology (such as machines); it includes nonmaterial techniques, which are either organizational or psychological, or both. Bureaucracy is an example of the former, whereas advertising and public relations are example of the latter.[7]

Technology increasingly dominates every form of human activity. In the West, technological innovations were integrated into the extant culture until the nineteenth century; that is, such innovations were situated in aesthetical, ethical, and religious relationships with other cultural artifacts and means of acting upon nature. Moreover, there were numerous limitations placed on the use of technology, both within and between societies.

5. Huxley, *Brave New World Revisited*, 25.

6. Huxley, *Brave New World Revisited*, 143.

7. Ellul, *Technological Society*.

The upshot of this is that prior to the nineteenth century technology was simply one aspect of a culture. This began to change when scientific and technological advances, and the concomitant "myth of progress," swept across Europe and North America in the nineteenth century. Science and technology were inexorably linked; indeed, technology as applied science became the justification for science. The incredible efflorescence of technological inventions bedazzled leaders and followers alike; consequently, technology became an end in itself, the purpose of civilization. As an end in itself, technology is simultaneously the most powerful means employed in the service of efficacy and efficiency. This desire to push technology as fast and as far as it will go demonstrates that technology, while a rational construction, is ultimately driven by the irrational will to power, the will to control, dominate, and exploit. Its material development was accompanied by its spiritualization. Technology was made sacred, and the sacred is that which is tacitly perceived to be of absolute power and absolute value.[8] The domination was now complete: technology was an uncontested material and spiritual power.

Technology becomes a system because we have looked for multiple users of the same technology, for example, laser technology, and have attempted to coordinate disparate technologies in the interest of efficiency. Technology is an open system in that it interacts with its two environments—nature and human society—but it is not open in that it does not possess genuine feedback. What finally allows technology to become an open system is the widespread use of the computer. The computer allows each technique to become a source of information for the coordination of the various technologies. Technology is a system, then, at the level of information. This means, however, that each subsystem loses some of its flexibility, for its courses of action must be adjusted to the needs of the other subsystems. The mutual interaction and mutual dependency of subsystems made possible by the computer is the technological system. In large urban areas the various technological subsystems such as communication, transportation, law enforcement, and commerce become more dependent on one another for the smooth operation of the overall urban system.

Although the technological system is an open system, it is more or less autonomous in relation to its human environment. The problem is that the technological system allows for no effective feedback, that is, self-regulation. Feedback means that a system (for instance, an ecological system) has the ability to correct the problem at its source. For example, if the technological system possessed feedback, then the use of the automobile, a major cause

8. Ellul, *New Demons*.

of air pollution, would be eliminated or severely curtailed. Instead, we attempt to discover ways of countering the negative effects of the automobile on the environment. We attempt to correct the problem after the fact so that we can have it both ways—drive our cars as much as we like and have a clean environment. Only humans, however, can provide feedback for the technological system. But because of our supreme faith in technology and because of our belief that technology itself can solve all problems, we do not perceive the need to provide such feedback.

Even if we attempt to use the computer as a feedback mechanism, it can only handle quantitative data. Hence, the computer rules out the possibility of evaluating the impact of technology on the qualitative side of life: how does technology affect culture and the human psyche? The ability of technology to create an efficient order at the societal level is offset by its disordering impact on culture and personality. The computer, however, is constitutionally unable to make such an historical and cultural interpretation.[9]

Technology's near-total domination (it affects us more than we affect it) is exemplified by the fact that today everything tends to be an imitation of technology and/or a compensation for its impact. Imitation of technology is nowhere more evident than in the glut of "how-to" books and techniques for relating to others. How to raise children, how to climb the ladder of success, how to manipulate one's boss, how to be popular, how to be happy—the list is unending. We can't help reducing everything to a technique, for technology deconstructs and supplants a common culture including common sense. As a consequence, almost everything today has to be learned as a technique. This is the paramount reason schools are forced to teach "life skills." Everything from babysitting to getting along with one's peers has to be learned as a formalized (technical) skill.

Technology creates the need for compensatory mechanisms in large part because of its impact on cultural meaning and the individual.[10] In traditional societies, practical knowledge was embedded in social institutions, which together formed the basis of a culture. Institutions contained what have been termed symbolically mediated experiences. Technology supplants experience and deconstructs common meaning. Cultural meaning, whether ethical or aesthetical, is now fragmented. This creates a desperate search for meaning, as with the proliferation of new religious groups and cults, or with the multiplication of business ethics, medical ethics, and so forth. Moreover, because there is an inverse relationship between power and values, technology, which today is exclusively about power—efficiency and

9. Ellul, *Technological System.*
10. Ellul, *Perspectives on Our Age,* 38–45.

efficacy—turns power itself into a value. Technology permits morality only insofar as it is reduced to ideology.

One of Ellul's most profound insights is that technology is concurrently the chief organizing force in modern society and its fundamental disorganizing force.[11] Technology supplants institutions and morality. At the same time, however, it leads to cultural and psychological fragmentation. We look to technology to repair the damage done to environment, culture, and psyche. The technological system, Ellul explains, creates and elaborates means of facilitation, adjustment, and compensation. Clearly humans make efforts to repair the damage technology does to the environment. For example, we recognize the need to reduce the release of greenhouse gases. Less obvious are the varied attempts to help humans adjust to the demands of technology. The mental health industry, including self-help groups, and the pharmaceutical industry play a key role here. As part of an overall adjustment, technology provides a plethora of compensations: mass media entertainment and consumerism of every sort. Consumerism and entertainment are our compensations for the increasing control technology exercises over us. The only important conformity today is conformity to technology. This is why the technological system can tolerate diverse lifestyles, moralities, cults, and other cultural expressions—they do not threaten the advancement of the system.[12]

Yet the technological system is out of control despite all the efforts at repair, facilitation, adjustment, and compensation. For it lacks true feedback. At most we can reject those technological choices that are obviously too risky economically, but we continue to wager our future on technological progress that makes life more dangerous and unpredictable.

What Huxley Got Right

In both *Brave New World* (1946) and *Brave New World Revisited* (1958), Huxley understood perfectly well that efficiency, order, and conformity are the paramount concerns in a technological civilization. Then too he brilliantly described how techniques for the manipulation and control of humans were to be the means to achieve this. Manipulation and control, he demonstrated, would center on the human desire for happiness (reduced to pleasure).

Drugs, consumerism, and absorption into the peer group were to be the chief means of achieving a permanent state of bliss. In his recent

11. Ellul, *Technological System.*
12. Ellul, *Technological System.*

book *Artificial Happiness*, Ronald Dworkin demonstrates how psychotropic drugs, exercise and dieting, alternative medicine, and spirituality are often used to achieve a temporary state of happiness that is hollow because it is not earned.[13] By this he means that real happiness results from achievements, most notably from moral commitment and responsibility to others, which were absent in *Brave New World*.

Huxley understood the growing demand for conformity, that is, "team players," despite all the rhetoric about diversity. This necessarily involves the loss of freedom and individuality.

In a stroke of genius he perceived that freedom would be redefined as happiness.

What Huxley Missed

Huxley did not have a concept of technology as a system. In *Brave New World* (1946), the various technologies are coordinated by the World Controllers. Therefore technological power is still principally in the hands of humans, not the technological system. He did not understand just how abstract power had become. The political and economic elite have greater access to technology, but the ultimate power resides in the technological system.

Equally important is his failure to recognize that the technological system leads to irrational attacks on the system. The more rational and controlling society becomes, the greater the need to escape the social order in distractions, compensations, or misplaced revolt. Huxley knew that the so-called sexual revolution was not going against traditional morality, but was a rebellion against the technological order.[14] The sexual revolution only leads to an even greater technological control of sexuality, such as the reduction of sex to technical performance. Nevertheless, Huxley failed to understand that not all compensations were only permitted and encouraged. Ellul's insight is that the technological system indirectly produces *spontaneous* attacks on the system without really threatening it.[15]

Finally, Huxley missed the technological transformation of social institutions. In *Brave New World*, marriage, the family, morality, and religion disappear. In a technological civilization, however, social institutions remain intact in form, but their content is radically altered. In traditional societies, social institutions were based on experiences that were given symbolic meaning. Technology supplants experience and fragments and

13. Dworkin, *Artificial Happiness*.
14. Huxley, *Brave New World*, rev. ed., xiii.
15. Ellul, *Technological System*.

sterilizes symbolic meaning. Raising children, for example, is increasingly a technical activity dictated by experts.

Criticism notwithstanding, Huxley's two books constitute a remarkable achievement, especially in light of how early (1931) he grasped the overweening importance of technology in the modern world. These two books command reflection on the nature and value of science fiction.

For at least two centuries writers and artists have debated the purpose of art and literature. For some the purpose is purely aesthetical; it is imaginative possibility. The artist or writer should be free to do whatever fulfills his or her creative urge. For others art and literature should have a moral purpose; it should criticize reality when it has abandoned its moral purpose.

Huxley already understood the major ideas of *Brave New World Revisited* (1958) when he wrote *Brave New World* in 1931. The latter work, a dystopia, was intended to alert readers to a real danger—the erosion of freedom. Most other science fiction writers, fascinated by technology, explore aesthetical possibility without much serious reflection on the nature of modern society. When science fiction is only about the future possibility of technology, it then can readily serve as compensation for a bleak reality. Moreover, it can make even a bleak future pleasurable at the moment. Ellul is undoubtedly right about this kind of science fiction. Because Huxley's science fiction serves a moral purpose and is grounded in an accurate critique of the modern world, it is an exception to the compensatory nature of most science fiction.

Chapter 13 _____

Technology and Terrorism in the Movie *Brazil*

Originally published in the *Bulletin of Science, Technology and Society* 26, no. 1 (2006) 20–22.

The 1985 movie *Brazil*, directed by Terry Gilliam, was prescient. It anticipated not only the proliferation of terrorist attacks and their representations in the media but also the deeper relationship that has developed between technology and terrorism. Many writers, such as Mark Juergensmeyer[1] and Martin Marty and R. Scott Appleby[2] have placed terrorism in the context of the political agenda of religious fundamentalism—control of society in the face of widespread immorality and corruption. Whether fundamentalist terrorist movements are more political or more religious is often the focus of debate. Jacques Ellul has argued that modern technology destabilizes traditional cultures, including their religions, and therefore indirectly instigates movements (like fundamentalism) that attempt to restore order.[3]

Hence, there is another context for the analysis of terrorism, one that *Brazil* develops with remarkable perspicacity. This other context, and the deeper one at that, is technology. The relationship between technology and terrorism is oblique, whereas the religious-political context is obvious: overt religious and/or political motivation for terrorism with political consequences.

I will begin with a brief description of technology and terrorism in the movie and then move to a sociological interpretation of the movie. I am not suggesting that the director necessarily had this interpretation in mind

1. Juergensmeyer, *Terrorism.*
2. Marty and Appleby, *Glory and the Power.*
3. Ellul, *What I Believe,* 125–27.

when he made the movie, but rather that a movie, like a text, is more or less open-ended in respect to its application to the world beyond the text.

The Movie *Brazil*

The movie opens with a bomb going off in a department store, but a television set within remains on with a news commentator talking about how many years the terrorists have been at it and how their occasional successes could be attributed to "luck." There are several terrorist attacks after this, one in a restaurant and one in a shopping complex, but one never sees any terrorists, only bombs blowing up and the indifference of bystanders. Even so, the number of real incidents is few. Terrorism for the most part exists in numerous media reports.

The theme of the movie, however, is not terrorism but suspected terrorism. The only terrorists we encounter are suspected terrorists. The first two are Tuttle, a heating engineer (played by Robert De Niro) and Jill, a young woman who lives in a flat above the Buttle family. Mr. Buttle is mistaken for Tuttle and is brutally arrested and taken away. Jill attempts to convince officials that they have the wrong man and consequently becomes a suspected terrorist. The protagonist of the story, Sam, falls in love with Jill and consequently becomes a suspected terrorist.

Sam has a job in the vast governmental bureaucracy and is content to stay where he is. The bureaucracy is set within a dystopia, one in which the government, bureaucracy, and technology exert near-total control over the citizens. The utopia (actually dystopia) is evidenced by the blending of time and the dislocation of space. Technologies of the past are interspersed with those of the present and future, for example, the typewriter—computer with a small television screen affixed to it. Past, present, and future are so entangled that the setting is no time and nowhere. Life has become monotonous, and the individuals in the story appear cynical, indifferent, and jaded. Sam's only escape from this nightmare is the world of daydreams and dreams—a fantasy world.

To discover Jill's identity, Sam accepts a promotion to the Department of Information Retrieval. Sam quickly runs afoul of his own department when he finagles Jill's dossier from Jack, his boss and old chum. At this moment, he becomes the principal suspected terrorist.

The Department of Information Retrieval has the responsibility of extracting information from citizens by any means, including torture. The threat of terrorism is the pretext for the government's totalitarian control of its citizens. Consequently, disobedience, even suspected disobedience, to the state is equated with terrorism. Terrorism is perceived to be so great a threat, really an infinite threat, that to be under suspicion is to be guilty. The political state is the true terrorist; as in the original meaning of the term (French Revolution), it instigates a reign of terror. At the end of the movie, Sam's boss and friend Jack tortures him to get information about terrorist activities until Jack becomes insane—a complete break with reality.

The movie *Brazil* takes technology as its twin theme. Technology as machinery has ample representation in the film, from robotic devices to instruments of torture. But the crucial technological device is bureaucracy, which is parodied by the endless number of regulations and paperwork. The beginning of the movie has the police breaking into the flat of a suspected terrorist; once they have subdued the wrong man, it later turns out, the paperwork ensues. The man's wife has to sign a form for her husband and a form indicating she has a receipt for him.

Max Weber recognized a century ago that bureaucracy was a kind of machine, a technology.[4] Lewis Mumford[5] and especially Jacques Ellul[6] expanded this insight. For Ellul, *technique* was the totality of rational methods (both material and nonmaterial) devoted exclusively to efficiency.[7] Technique, at least in the modern world, is about power and organization and, hence, demands that nature and society be brought in line with its development. What becomes most important in a civilization that tacitly regards technology as its fate is the coordination of the disparate technologies: machine, bureaucracy, and propaganda (including advertising and public relations) are increasingly brought together to form a system. This theme is evident in *Brazil*, wherein the political state (itself technical in orientation) has integrated machinery, bureaucracy, and psychological techniques. Politics today has become thoroughly technical in its dependence on bureaucracy and the mass media; power has thus become abstract and complex and beyond full human control.

Interpretation

If both technology and terrorism are themes of *Brazil*, what is their connection? I wish to develop two ideas: (a) both are forms of violence and (b) terrorism is now the way the technological system perfects itself. I believe

---

4. Weber, *Economy and Society*, vol. 2.

5. Mumford, *Technics and Human Development*.

6. Ellul, *Technological Society*.

7. Ellul, *Technological Society*, xxv.

that the first idea was consciously developed in the movie whereas the second was only intimated.

Following the work of Ellul on technology, I maintain that technology is first and foremost about power. It represents the most powerful (effective and/or efficient) means of acting. But as power grows, moral values necessarily lose their efficacy. All moral values involve some limitation on power. For example, justice demands that I not take whatever I like but only that which is due me. Love requires that I not manipulate or force my will on another. At a certain threshold, power itself becomes a "value." Power, however, is a jealous value; it permits no other.

Moreover, any society dominated by technology is in the process of becoming totalitarian. Once technology becomes a system (the widespread and purposeful coordination of technologies), it likewise becomes our milieu, an environment both material and symbolic. The irony of this is that technology, our creation, has become our only possibility to resolve all issues, natural and social. As Ellul notes, when every possibility is technological, technology becomes necessity.[8] As a result, human freedom vanishes. The new totalitarianism is technological before it is political or religious.

Walter Lacquer has demonstrated that terrorists are primarily interested in the acquisition of power.[9] He cites terrorists who compare the blowing up of a bomb to an orgasm. In his view, religious and political ideology is primarily a screen for the real terrorist interest in power. Terrorism thus mimics the power and totalitarianism of technology. This helps explain the widespread presence of violence in the media. Violence, including terrorism in the media, is the reverse image of technology. Technology is abstract and rational, violence is concrete and irrational; both are expressions of ultimate power.

Terrorism reinforces the technological system even when it appears to go against the system. Terrorism is directed against the political state or at least political, religious, and economic power. But terrorists who employ technology, including the media, for their purposes are at one with those in power in embracing technology. What terrorists actually accomplish is the continued development of the technological system. They point out the weaknesses of the system as it is institutionalized in society. They force societies to improve, among other things, their military technology, intelligence-gathering techniques, computer technology, and propaganda. Each terrorist innovation has to be countered by a technological advance. At a certain

8. Ellul, *Technological Bluff*, 217–20.
9. Lacquer, *Age of Terrorism*.

point, there is no difference between terrorists and those in power; both employ whatever violence and technology are at hand.

This is why, I think, one never sees a terrorist in *Brazil*. The only visible terrorist is the government pursuing suspected terrorists, which makes one wonder if the government performed the bombings to justify its totalitarian onslaught against suspected terrorism. Everyone is either a terrorist or a suspected terrorist.

A third theme of freedom appears in *Brazil* and is set against the technology-terrorism relationship. In the film, Sam initially finds freedom in his dreams and daydreams. Freedom here is escape. After falling in love with Jill, he asserts freedom by going against the bureaucracy in his attempt to clear her name. Sam becomes aligned with Tuttle, the heating engineer, who is perceived to be a terrorist because he repairs heating and cooling systems without the necessary paperwork. When Sam is finally apprehended and tortured, he finds freedom in insanity, the ultimate escape. In *The Technological Society*, Ellul observes that insanity is perhaps the only way to avoid the technological juggernaut.[10] His hyperbole was meant to awaken the reader to the danger to freedom that technology presents. I think that *Brazil*, as well, is giving us a warning. Rather than saying, "All is hopeless," the message is, "Act before it is too late."

Brazil's Lesson for Today

Behind the political rhetoric on both sides—government and terrorist—the real victor is technology, which does not care whether right or left, religious or atheist, terrorist or conformist has power. For all are united in their support of technology and its apparently inevitable advance.

---

10. Ellul, *Technological Society*, 410–27.

Bibliography

Adorno, Theodor, et al. *The Authoritarian Personality*. New York: Norton, 1969.

Allport, Gordon. *The Nature of Prejudice*. Garden City, NY: Anchor, 1958.

Angle, Paul. *Bloody Williamson*. New York: Knopf, 1952.

Arendt, Hannah. *Eichmann in Jerusalem*. New York: Viking, 1963.

Barfield, Owen. *Poetic Diction*. New York: McGraw-Hill, 1964.

————. *Saving the Appearances*. New York: Harcourt Brace Jovanovich, 1957.

Barnow, Victor. *Cultural Anthropology*. New York: McGraw-Hill, 1991.

Baudrillard, Jean. "Consumer Society." In *Jean Baudrillard: Selected Writings*, edited by Mark Poster, 29–56. Stanford, CA: Stanford University Press, 1988.

————. *Simulations*. Translated by Paul Foss, Paul Patton, and Philip Beitchman. New York: Semiotext(e), 1983.

Baum, Andrew, Jerome Singer, and Carlene Baum. "Stress and the Environment." In *Environmental Stress*, edited by Gary Evans, 15–44. New York: Cambridge University Press, 1982.

Benjamin, Walter. "The Work of Art in the Age of Mechanical Reproduction." In *Illuminations*, translated by Harry Zohn, 217–51. New York: Schocken, 1969.

Bergson, Henri. *The Two Sources of Morality and Religion*. Translated by R. Ashley Audra and Cloudesley Brereton. Garden City, NY: Anchor, 1954.

Bird-David. "Beyond 'The Original Affluent Society': A Culturalist Reformulation." *Current Anthropology* 33 (1992) 25–34.

Blumberg, Abraham. *Criminal Justice*. Chicago: Quadrangle, 1970.

Boorstin, Daniel. *The Americans: The Democratic Experience*. New York: Random House, 1973.

————. *Democracy and Its Discontents*. New York: Random House, 1974.

————. *The Image*. New York: Harper and Row, 1961.

Bowers, C. A. *The Cultural Dimension of Educational Computing*. New York: Teachers College, 1988.

Bradley, Gunilla. *Computers and the Psychosocial Work Environment*. London: Taylor and Francis, 1989.

Brod, Craig. *Technostress*. Reading, MA: Addison-Wesley, 1984.

Brooks, David. "The Organizational Kid." *Atlantic* 290 (2001) 40–54.

Brown, Lauren, and J. E. Cima. "The Illinois Chorus Frog and the Sand Lake Dilemma." In *The Status and Conservation of Midwestern Amphibians*, edited by Michael Lannoo, 301–11. Iowa City, IA: University of Iowa Press, 1998.

Burkert, Walter. *Homo Necans*. Translated by Peter Bing. Berkeley, CA: University of California Press, 1983.

———. "The Problem of Ritual Killing." In *Violent Origins*, edited by R. Hamerton-Kelly, 141–90. Stanford, CA: Stanford University Press, 1987.

Caillois, Roger. *Man, Play and Games*. Translated by Meyer Barash. Urbana, IL: University of Illinois Press, 2001.

———. *Man and the Sacred*. Translated by Meyer Barash. Westport, CT: Greenwood, 1980.

Canetti, Elias. *Crowds and Power*. Translated by Carol Stewart. New York: Seabury, 1978.

Carpenter, Edmund. *Oh, What a Blow that Phantom Gave Me*. New York: Holt, Rinehart and Winston, 1973.

Carr, Nicholas. *The Shallows*. New York: Norton, 2010.

Cassirer, Ernest. *Language and Myth*. Translated by Susanne Langer. New York: Dover, 1953.

Clark, G. R. *Mouth of the Mahomet Regional Groundwater Model, Imperial Valley Region of Mason, Tazewell and Logan Counties, Illinois*. Springfield, IL: Illinois Department of Transportation, 1995.

Crosby, Alfred. *The Measure of Reality*. New York: Cambridge University Press, 1997.

Cross, Gary. *Men to Boys*. New York: Columbia University Press, 2008.

Debord, Guy. *Society of the Spectacle*. Detroit: Black and Red, 1977.

Douglas, Ann. *The Feminization of American Culture*. New York: Anchor, 1988.

Douglas, Mary. *Natural Symbols*. New York: Vintage, 1973.

Dumochel, Paul. "Introduction." In *Violence and Truth*, edited by Paul Dumochel, 1–22. Stanford, CA: Stanford University Press, 1985.

Dumont, Louis. "The Anthropological Community and Ideology." In *Essays on Individualism*, 202–33. Chicago: University of Chicago Press, 1986.

———. *From Mandeville to Marx*. Chicago: University of Chicago Press, 1977.

———. *Homo Hierarchicus*. Translated by Mark Sainsbury, Louis Dumont, and Basia Gulati. Chicago: University of Chicago Press, 1980.

———. "On Value, Modern and Nonmodern." In *Essays on Individualism*, 234–68. Chicago: University of Chicago Press, 1986.

Dupuy, Jean-Pierre. "Myths of the Informational Society." In *The Myths of Information*, edited by Kathleen Woodward, 3–17. Madison: Coda, 1980.

———. *On the Origins of Cognitive Science*. Translated by M. B. De Bevoise. Cambridge, MA: MIT Press, 2009.

Durkheim, Emile. *Suicide*. Translated by John Spaulding and George Simpson. New York: Free, 1951.

Dworkin, Ronald. *Artificial Happiness*. New York: Carroll and Graff, 2006.

Ekirch, Arthur. *The Decline of American Liberalism*. New York: Atheneum, 1980.

Eliade, Mircea. *The Myth of the Eternal Return*. Translated by Willard Trask. Princeton, NJ: Princeton University Press, 1965.

———. *Patterns in Comparative Religion*. Translated by Rosemary Sheed. New York: New American Library, 1974.

———. *The Quest*. Chicago: University of Chicago Press, 1969.

———. *The Sacred and the Profane*. Translated by Willard Trask. New York: Harper and Row, 1961.

Elias, Norbert. *The History of Manners*. Translated by Edmund Jephcott. New York: Pantheon, 1978.

Ellul, Jacques. *The Betrayal of the West*. Translated by Matthew O'Connell. New York: Seabury, 1978.

———. *A Critique of the New Commonplaces*. Translated by Helen Weaver. New York: Knopf, 1968.

———. *The Ethics of Freedom*. Translated by Geoffrey Bromiley. Grand Rapids: Eerdmans, 1976.

———. *Hope in Time of Abandonment*. Translated by C. Edward Hopkin. New York: Seabury, 1973.

———. *The Humiliation of the Word*. Translated by Joyce Hanks. Grand Rapids: Eerdmans, 1985.

———. *If You Are the Son of God*. Translated by Anne-Marie Andreasson-Hogg. Eugene, OR: Cascade, 2014.

———. "Life Has Meaning." In *What I Believe*, translated by Geoffrey Bromiley, 13–18. Grand Rapids: Eerdmans, 1990.

———. *The New Demons*. Translated by C. Edward Hopkin. New York: Seabury, 1975.

———. *Perspectives on Our Age*. Edited by Willem Vanderburg. Translated by Joachim Neugroschel. Toronto: CBC Enterprises, 1981.

———. *The Political Illusion*. Translated by Konrad Kellen. New York: Vintage, 1967.

———. *Propaganda*. Translated by Konrad Kellen. New York: Vintage, 1969.

———. "Symbolic Function, Technology and Society." *Journal of Social and Biologic Structure* 1 (1978) 207–18.

———. *The Technological Bluff*. Translated by Geoffrey Bromiley. Grand Rapids: Eerdmans, 1990.

———. *The Technological Society*. Translated by John Wilkinson. New York: Knopf, 1964.

———. *The Technological System*. Translated by Joachim Neugroschel. New York: Continuum, 1980.

———. "Toward a Forum-Style University." In *In Season Out of Season*, translated by Lani Niles, 158–71. New York: Harper and Row, 1982.

———. *To Will and To Do*. Translated by C. Edward Hopkin. Philadelphia: Pilgrim, 1969.

———. *What I Believe*. Translated by Geoffrey Bromiley. Grand Rapids: Eerdmans, 1989.

Erikson, Erik. *Identity: Youth and Crisis*. New York: Norton, 1968.

Evans, Gary, ed. *Environmental Stress*. New York: Cambridge University Press, 1982.

Ewen, Stuart. *All Consuming Images*. New York: Basic, 1988.

Eyal, Nir. *Hooked*. New York: Penguin, 2014.

Friedenberg, Edgar. *The Vanishing Adolescent*. New York: Dell, 1959.

Friedman, Lawrence. *The Horizontal Society*. New Haven, CT: Yale University Press, 1999.

Furet, Francois. "The Conceptual System of 'Democracy in America.'" In *In the Workshop of History*, translated by Jonathan Mandelbaum, 167–96. Chicago: University of Chicago Press, 1984.

Garfinkel, Harold. *Studies in Ethnomethodology*. Englewood Cliffs, NJ: Prentice-Hall, 1967.

Geertz, Clifford. *The Interpretation of Cultures*. New York: Basic, 1973.

Gehlen, Arnold. *Man in the Age of Technology*. Translated by Patricia Lipscomb. New York: Columbia University Press, 1980.

Gerbner, George, and Larry Gross. "Living with Television." *Journal of Communication* 26.2 (1976) 173–97.

Gibson, James William. *Warrior Dreams*. New York: Hill and Wang, 1994.

Gilliam, Terry, director. *Brazil*. Embassy International Pictures and Universal Pictures, 1985.

Girard, Rene. "Generative Scapegoating." In *Violent Origins*, edited by Robert Hamerton-Kelly, 73–105. Stanford, CA: Stanford University Press, 1987.

———. *The Scapegoat*. Translated by Yvonne Freccero. Baltimore: Johns Hopkins University Press, 1986.

———. *Things Hidden Since the Foundation of the World*. Translated by Stephen Bann and Michael Metteer. Stanford, CA: Stanford University Press, 1987.

———. *Violence and the Sacred*. Translated by Patrick Gregory. Baltimore: Johns Hopkins University Press, 1977.

Gitlin, Todd. *Media Unlimited*. New York: Metropolitan, 2001.

Gleick, James. *Faster*. New York: Pantheon, 1999.

Goldberger, Paul. "Design: the Risks of Razzle-Dazzle." *New York Times*, April 12, 1987.

Gombrich, E. H. "The Visual Image." *Scientific American* 227 (September, 1972) 82–96.

Gottlieb, Beatrice. *The Family in the Western World*. New York: Oxford University Press, 1993.

Graham, Paul. "The Acceleration of Addictiveness." Accessed November 12, 2013, www.paulgraham.com/addiction.html.

Hayles, N. Katherine. *How We Became Posthuman*. Chicago: University of Chicago Press, 1999.

Healy, Jane. *Endangered Minds*. New York: Touchstone, 1991.

Heider, Fritz. *The Psychology of Interpersonal Relations*. New York: Wiley, 1958.

Hendin, Herbert. *The Age of Sensation*. New York: Norton, 1975.

Herberg, Will. *Protestant, Catholic, Jew*. Chicago: University of Chicago Press, 1973.

Hochschild, Arlie. *The Managed Heart*. Berkeley, CA: University of California Press, 1983.

———. *The Time Bind*. New York: Metropolitan, 1997.

Horkheimer, Max, and Theodor Adorno. *Dialectic of Enlightenment*. Translated by John Cumming. New York: Herder and Herder, 1972.

Horney, Karen. *The Neurotic Personality of Our Time*. New York: Norton, 1937.

Hudson, Kenneth. *The Jargon of the Professions*. London: Macmillan, 1978.

———. *The Language of the Teenage Revolution*. London: Macmillan, 1983.

Huizinga, J. H. *In the Shadow of Tomorrow*. New York: Norton, 1936.

———. *The Waning of the Middle Ages*. Translated by Frederik Hopman. Garden City, NY: Anchor, 1954.

Hummel, Ralph. *The Bureaucratic Experience*. New York: St. Martin's, 1982.

Hunter, James Davison. *The Death of Character*. New York: Basic, 2000.

Huxley, Aldous. *Brave New World*. London: Chatto and Windus, 1932.

———. *Brave New World*. Rev. ed. New York: Harper and Row, 1946.

———. *Brave New World Revisited*. New York: Harper and Row, 1958.

Illich, Ivan. *Disabling Professions*. Edited by Ivan Illich, et al. Boston: Marion Boyars, 1978.

Jackall, Robert. *Moral Mazes*. New York: Oxford University Press, 1988.

Jackson, Maggie. *Distracted*. Amherst, NY: Prometheus, 2009.

Jensen, Adolf. *Myth and Cult Among Primitive Peoples*. Translated by Marianna Cholden and Wolfgang Weissleder. Chicago: University of Chicago Press, 1963.

Jouvenel, Bertrand de. *On Power*. Translated by J. F. Huntington. Boston: Beacon, 1962.

Juergensmeyer, Mark. *Terrorism in the Mind of God*. Berkeley, CA: University of California Press, 2000.

Kahler, Erik. *The Tower and the Abyss*. New York: Braziller, 1957.

Kelsen, Hans. *Society and Nature*. New York: Arno, 1974.

Kerr, Walter. *The Decline of Pleasure*. New York: Simon and Schuster, 1965.

Kierkegaard, Søren. *Concluding Unscientific Postscript*. Translated by Howard Hong and Edna Hong. Princeton: Princeton University Press, 1992.

———. *The Present Age*. Translated by Alexander Dru. New York: Harper and Row, 1962.

———. *The Sickness Unto Death*. Translated by Howard Hong and Edna Hong. Princeton, NJ: Princeton University Press, 1980.

Kottak, Conrad. *Cultural Anthropology*. New York: McGraw-Hill, 1991.

Kundera, Milan. *The Art of the Novel*. Translated by Leslie Asher. New York: Grove, 1988.

———. *Slowness*. New York: Harper Collins, 1996.

Lacquer, Walter. *The Age of Terrorism*. Boston: Little, Brown, 1987.

Leibovitz, Liel. *God in the Machine*. West Conshohocken, PA: Templeton, 2013.

Leroux, Charles. "Our Electronic Friends." *Chicago Tribune*, May 6, 2001.

Levine, Robert. *A Geography of Time*. New York: Basic, 1997.

Lewontin, Richard. *Biology as Ideology*. New York: Harper Collins, 1992.

———. *The Triple Helix*. Cambridge, MA: Harvard University Press, 2001.

Linder, Staffan. *The Harried Leisure Class*. New York: Columbia University Press, 1970.

Littleton, C. Scott. *The New Comparative Mythology*. Berkeley, CA: University of California Press, 1973.

MacIntyre, Alasdair. *After Virtue*. Notre Dame: University of Notre Dame Press, 1984.

Mack, Burton. "Religion and Ritual." *In Violent Origins*, edited by Robert Hamerton-Kelly, 1–72. Stanford, CA: Stanford University Press, 1987.

Mander, Jerry. *Four Arguments for the Elimination of Television*. New York: Quill, 1978.

———. *In the Absence of the Sacred*. San Francisco: Sierra Club Books, 1991.

Maringer, Johannes. *The Gods of Prehistoric Man*. Translated by Mary Ilford. New York: Knopf, 1960.

Marshack, Alexander. *The Roots of Civilization*. New York: McGraw-Hill, 1972.

Martin, Jay. *Who Am I This Time?* New York: Norton, 1988.

Marty, Martin E., and R. Scott Appleby. *The Glory and the Power*. Boston: Beacon, 1992.

Marx, Karl. *The Economic and Philosophic Manuscripts of 1844*. Translated by Martin Milligan. New York: International, 1964.

Miller, Mark. *Boxed In*. Evanston, IL: Northwestern University Press, 1988.

Mumford, Louis. *Technics and Human Development*. New York: Harcourt Brace Jovanovich, 1967.

Musil, Robert. *The Man Without Qualities*. Translated by Sophie Wilkins and Burton Pike. New York: Knopf, 1995.

"Old Stories about Negro Lake." *Mason County Democrat*, September 15, 1993.

Ong, Walter. *Orality and Literacy*. New York: Metheun, 1982.

O'Reilly, Charles. "Variations in Decision Makers' Use of Information Sources." *Academy of Management Journal* 25 (1982) 756–71.

Orwell, George. *1984*. New York: New American Library, 1949.

Otto, Rudolph. *The Idea of the Holy*. Translated by John Harvey. New York: Oxford University Press, 1958.

Parmalee, P. W., and F. D. Loomis. *Decoys and Decoy Carvers of Illinois*. DeKalb, IL: Northern Illinois University Press, 1969.

Pattison, Robert. *The Triumph of Vulgarity*. New York: Oxford University Press, 1987.

Postman, Neil. *Amusing Ourselves to Death*. New York: Viking, 1985.

————. "The News." In *Conscientious Objections*, 72–81. New York: Knopf, 1988.

————. "The Parable of the Ring Around the Collar." In *Conscientious Objections*, 66–81. New York: Knopf, 1988.

Renfrew, Colin. *Archeology and Language*. New York: Cambridge University Press, 1987.

Ricoeur, Paul. *Interpretation Theory*. Fort Worth, TX: Texas Christian University Press, 1976.

————. *Lectures on Ideology and Utopia*. Edited by George Taylor. New York: Columbia University Press, 1986.

Riesman, David. *The Lonely Crowd*. New Haven, CT: Yale University Press, 1950.

Rifkin, Jeromy. *The End of Work*. New York: Putnam, 1995.

Sahlins, Marshall. *Stone Age Economics*. Chicago: Aldine-Atherton, 1972.

Sapir, Edward. "Culture, Genuine and Spurious." In *Culture, Language and Personality*, edited by David Mandelbaum, 78–119. Berkeley, CA: University of California Press, 1970.

Schivelbusch, Wolfgang. *The Railway Journey*. Berkeley, CA: University of California Press, 1986.

Schüll, Natasha. *Addiction by Design*. Princeton, NJ: Princeton University Press, 2012.

Simmel, Georg. "The Metropolis and Mental Life." In *The Sociology of Georg Simmel*, translated by Kurt Wolff, 409–24. New York: Free, 1950.

Smith, Jonathan Z. *To Take Place*. Chicago: University of Chicago Press, 1987.

————. "The Wobbling Pivot." In *Map Is Not Territory*, 89–128. Leiden: E. J. Brill, 1978.

Sommers, Christina. "Ethics Without Virtue." *The American Scholar* 53 (Summer, 1984) 381–89.

Speckman, F. C. "Facts about 'Nigger' Lake." *Mason County Democrat*, March 3, 1960.

Stivers, Richard. *The Culture of Cynicism*. Cambridge, MA: Blackwell, 1994.

————. "The Deconstruction of the University." *The Centennial Review* 35 (1991) 115–36.

————. *Evil in Modern Myth and Ritual*. Athens, GA: University of Georgia Press, 1982.

————. "The Festival in Light of the Theory of the Three Milieus: A Critique of Girard's Theory of Ritual Scapegoating." *Journal of the American Academy of Religion* 61.3 (Autumn, 1993) 505–538.

————. *The Illusion of Freedom and Equality*. Albany, NY: SUNY Press, 2008.

————. *Shades of Loneliness*. Lanham, MD: Rowman and Littlefield, 2004.

————. *Technology as Magic*. New York: Continuum, 1999.

Szasz, Thomas. *The Manufacture of Madness*. New York: Harper and Row, 1970.

Talbott, Stephen. *The Future Does Not Compute*. Sebastopol: O'Reilly, 1995.

Tocqueville, Alexis de. *Democracy in America*. Translated by George Lawrence. Garden City, NY: Anchor, 1969.

Todorov, Tzvetan. *Facing the Extreme*. Translated by Arthur Denner and Abigail Pollak. New York: Metropolitan, 1996.

———. "Notes from Underground." In *Genres in Discourse*, translated by Catherine Porter, 72–92. Cambridge: Cambridge University Press, 1990.

———. "Some Remarks on Contacts among Cultures." In *The Morals of History*, translated by Alyson Waters, 71–84. Minneapolis: University of Minnesota Press, 1995.

Trow, George. *Within the Context of No Context*. New York: Atlanta Monthly, 1981.

Tudor, Andrew. *Image and Influence*. New York: St. Martin's, 1975.

Turkle, Sherry. *Alone Together*. New York: Basic, 2011.

Turnbull, Colin. *The Forest People*. New York: Simon and Schuster, 1961.

Turner, Victor. *The Ritual Process*. Ithaca, NY: Cornell University Press, 1977.

Ucko, Peter, and Andree Rosenfeld. *Paleolithic Cave Art*. New York: McGraw-Hill, 1967.

Vahanian, Gabriel. *The Death of God*. New York: Braziller, 1961.

van den Berg, J. H. *The Changing Nature of Man*. Translated by H. F. Croes. New York: Norton, 1961.

———. *Divided Existence and Complex Society*. Pittsburgh: Duquesne University Press, 1974.

———. *Medical Power and Medical Ethics*. New York: Norton, 1978.

———. "What is Psychotherapy?" *Humanitas* 7.3 (1971) 322–70.

Vanderburg, Willem. *The Growth of Minds and Cultures*. Toronto: University of Toronto Press, 1985.

———. *The Labyrinth of Technology*. Toronto: University of Toronto Press, 2000.

Wallace, Anthony. *The Death and Rebirth of the Seneca*. New York: Vintage, 1969.

Weber, Max. *Economy and Society*. Vol. 2. Translated by Ephrains Fischoff, et al. Berkeley, CA: University of California Press, 1978.

Wheelwright, Philip. *Metaphor and Reality*. Bloomington, IN: Indiana University Press, 1962.

White, Mimi. *Tele-Advising*. Chapel Hill, NC: University of North Carolina Press, 1992.

Whyte, William. *The Organization Man*. New York: Anchor, 1956.

Williams, D. "Ditch Proposed to Slow Flooding." *Peoria Journal Star*, December 2, 1993.

Williams, Tannis. *The Impact of Television*. Orlando: Academic, 1986.

Wolfenstein, Martha. "The Emergence of Fun Morality." *Journal of Social Issues* 7 (1951) 15–25.

Zuboff, Shoshana. *In the Age of the Smart Machine*. New York: Basic, 1988.